D1084689

Plastics Processing Technology

Edward A. Muccio

Manager, Book Acquisitions— Veronica Flint

Production Project Manager—Suzanne E. Frueh

Production Project Coordinator—Nancy M. Sobie

**The Materials
Information Society**

First printing, May 1994
Second printing, September 1994
Third printing, January 1997

This book is a collective effort involving hundreds of technical specialists. It brings together a wealth of information from worldwide sources to help scientists, engineers, and technicians solve current and longrange problems.

Comments, criticisms, and suggestions are invited, and should be forwarded to ASM International.

Library of Congress Cataloging-in-Publication Data
Plastics processing technology / Edward A. Muccio
    p. cm.
Includes bibliographical references and index
ISBN: 0-87170-494-3
SAN: 204-7586
1. Plastics. I. Title
TP1120.M793 1994   94-15959
668.4'1—dc20   CIP

ASM International®
Materials Park, OH 44073-0002

Printed in the United States of America

Acknowledgments

I would like to acknowledge the time, support, and advice provided by the individuals who offered to review chapters of this book:

- Paul Bugajski, NIST/ Midwest Manufacturing Technology Center
- David H. Harry, Iowa Plastics Technology Center
- Dan Hochgreve and Munjal Parikh, Industrial Technology Institute
- Erik Lokensgard, Eastern Michigan University
- John McLeod, Thoreson-McCosh
- Frank R. Nissel, Welex Incorporated
- Fyodor (Ted) Shutov, Center of Excellence in Polymer Science and Engineering, Illinois State University

I would also like to thank the people who contributed to the text:

- The employees of Watlow Electric Co. for the material about temperature control (Chapter 3)

- Jack St. Pierre, Pocono P.E.T., for his significant contribution to Chapter 5

- C. D. Shirrell, Shell Development Center, for a significant portion of the material about structural reaction injection molding (Chapter 7)

- Richard K. Okine, Engineering Research and Development Division, E.I. du Pont de Nemours & Co., Inc., for the material about thermoplastic composite sheet forming, including the tables (Chapter 7)

- Earl M. Zion, EMZ Associates, for the material about contact molding, pultrusion, and filament winding, including Table 7-6 (Chapter 7)

- James E. Snyder, Polymer Design Corporation, for the material about resin casting (Chapter 8)

- The employees of Carpco, Inc. for the material about separating plastics for recycling (Chapter 8)

- John McLeod, Thoreson-McCosh, Inc., for the material about drying (Chapter 9)

Finally, I would like to express my thanks to the students of the Ferris State University Plastics Engineering Technology program for their interest and enthusiasm. A special note of thanks to two of my students, Steve Gay and Peggy Childress, for the time and energy they spent developing some of the graphics and illustrations.

Preface

This book was developed and written to provide the reader with a basic understanding of plastics processing technology. The assumption is made that the reader has little or no experience in plastics product manufacturing or plastics processing. The book is written at a level that should be helpful to technicians, managers, buyers, quality assurance personnel, and engineers who have minimal experience with plastics.

Plastics processing is no longer the sole domain of custom plastics manufacturers. It is now a significant component of other industries. Dairies and drink processors now mold their own bottles, electronic manufacturers now produce their own plastic components and packaging, and automobile manufacturers control inventory, quality, and cost by producing the ever-increasing amounts of plastic parts for cars and trucks.

Plastics processing is a complex subject that involves a thorough understanding of materials, thermodynamics, fluid technology, control, and tool/part design. This book will highlight the key aspects of these areas, using clear and simple illustrations to provide the reader with an awareness of plastics processing technology.

Dedication

I would like to dedicate this book to a man who has shown a strength and love beyond which even he thought was possible.

A man I love and am proud to call my father, Edward Muccio.

Modern injection molding facility

Table of Contents

1

Introduction: Assessing Plastics Processors

Plastics processing is one of the fastest-growing segments of manufacturing. Plastic materials, as measured in volume, have now exceeded steel in terms of processed quantity. Furthermore, the growth of plastics processing is not isolated to the United States; it is a global phenomenon.

The increasing development of plastics processing is directly related to the desire of consumers to have products produced with plastic materials. Many of today's products (e.g., VCRs, CDs, and high-performance vehicles) can be manufactured only with plastic materials.

Globally, there are three main plastics processing areas: North America (the United States and Canada); the Pacific Rim (notably Japan, Taiwan, Singapore, China, and Korea); and Europe (notably Germany, Austria, Italy, France, and Switzerland). There are, however, growth areas that will become significant plastics processing locations over the next 20 years. These areas are Mexico, Brazil, Eastern Europe, South Africa, India, and the former Soviet Union.

What Is Plastics Processing?

Plastics processing is a form conversion process. Plastic part manufacturers purchase plastic materials in the form of plastic powder, pellets, or beads (Fig.

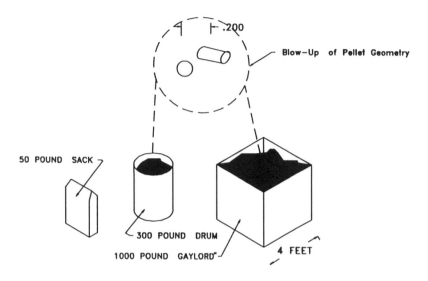

Fig. 1-1 Plastic containers and pellet shape

Fig. 1-2 Plastic part produced by injection molding. Courtesy of Miles, Inc.

Fig. 1-3 Example of plastics processing tooling. This is an injection molding machine. Courtesy of Husky Injection Molding Systems

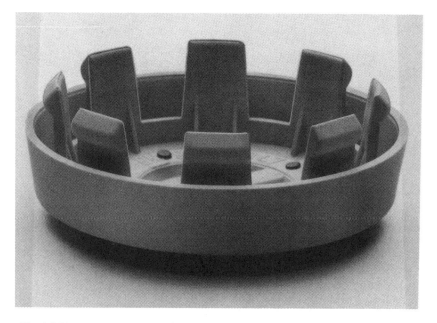

Fig. 1-4 Key growth markets for plastic parts

1-1). These are heated, melted, and transformed into usable plastic products (Fig. 1-2) by the plastics processing equipment and tooling (Fig. 1-3). These processes are growing in terms of the sophistication of control and the complexity of the types of products made. An example of the complexity of plastic products can be seen in Fig. 1-4.

Who Are the Plastics Processors?

Some common attributes of plastics processors (plastics manufacturers) are that they:

- Produce products (all or in part) from plastic materials
- Do not manufacture the basic plastic material, but instead "convert" the plastic resin into plastic parts by using plastics processing equipment
- Have the same 15 major areas of responsibility, as discussed in the section "Responsibilities of Plastics Processors" in this chapter

Plastics processors can be categorized as either captive or custom (Fig. 1-5).

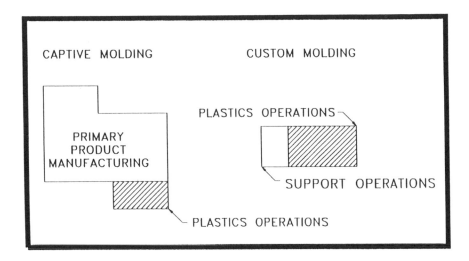

Fig. 1-5 Differences between captive molding and custom molding

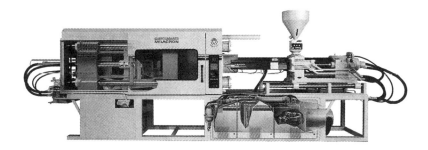

Fig. 1-6 Example of the use of plastic in electronics manufacturing. Courtesy of Husky Injection Molding Systems

Captive plastics processors are usually components of larger manufacturing operations (e.g., automaking, dairies, furniture manufacturing, sporting goods manufacturing, building products manufacturing, and consumer electronics manufacturing, as in Fig. 1-6). The plastic parts they produce are typically used in subsequent assembly operations. Captive processors usually exist to maximize delivery and minimize cost. They cost plastic products without a profit margin. Disadvantages may include limited technical skills, scheduling problems, and high workforce turnover.

It is often very difficult to determine which manufacturers have captive plastics processing capabilities, because there is no need to advertise or communicate such a capability to the public. These captive plastics processors are literally "hidden plastics manufacturers."

Custom plastics processors are "job shops" that produce plastic products for a broad range of customers. Their strengths are their technical specialists, diverse experience base, and strong technical support. They often have the capability to design, tool, and manufacture plastic parts. Custom plastics processors set prices that include a profit margin, and they must advertise their existence to attract new business. Other disadvantages may include rapid schedule fluctuations and shipping costs.

Trends In Processing

Over the past 100 years, plastics processing has evolved from crude equipment, which struggled to melt the material and force material through a die or into a mold, to the current microprocessor-controlled plastics processing equipment (Fig. 1-7). The future of plastics processing will be in the areas of control logic, integration with other processes, rapid tooling/ materials changeover, and automation.

The in-place equipment base throughout the plastics industry varies, because well-maintained equipment can easily last over 20 years. It is possible, therefore, to have conventional relay-logic-controlled equipment and computer-controlled equipment in the same plastics manufacturing plant. This diversity of equipment can cause confusion for manufacturing personnel if they are required to operate and maintain both conventional and microprocessor-based technologies.

The "technology stretch" can be compared to the situation of a child who learned to tell time only on a digital watch. Certain concepts may not have been learned, such as the ability to visualize time other than in the present. Children who learn to tell time using a digital watch may thus have problems telling time using an analog watch (one with hands), and vice versa. Similarly, using a video display terminal to learn how to use a computer-controlled plastics processing machine may severely limit the troubleshooting capabilities of the operator, because there is no opportunity to see the cause and effect of what is being controlled. Regardless of how sophisticated plastics processing equipment becomes, it is important that manufacturing personnel understand the basic concepts of each process.

Fig. 1-7 Example of microprocessor-controlled plastics processing equipment. This is an injection molding machine. Courtesy of Cincinnati Milacron

Because this book is intended for use by individuals who are not educated specifically in plastics technology, it will be useful to explore what factors should be considered when assessing a plastics processor. This is an important step in obtaining quality plastic parts. Assessments are performed by supplier quality improvement programs (SQIP), by other plastics processors who want to benchmark their own operations, by buyers and quality engineers, and by entrepreneurs who need to determine whether an idea should be converted to an actual product. The following guidelines may be helpful for both plastics processors and customers of plastics processors.

Responsibilities of Plastics Processors

1. Product Design Review

Successful production of plastic parts requires that the product design be reviewed by the processor. The plastics manufacturing process has a greater

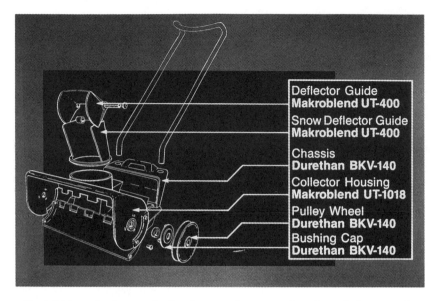

Fig. 1-8 Example of an application that uses plastic parts but whose designer may be more familiar with conventional materials. Courtesy of Miles, Inc.

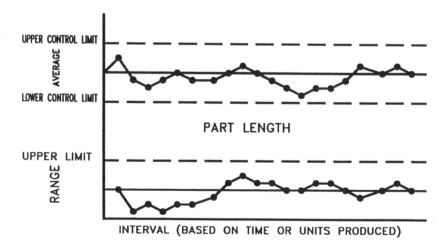

Fig. 1-9 Statistical process control chart

effect on the quality of a plastic part than on the quality of parts manufactured using conventional materials (e.g., wood, glass, or metal). The design of a plastic part also has a significant effect on its manufacturability. Details such as wall thickness, respect for design rules, and shrinkage need to be considered by the part designer. Many plastic parts are designed by individuals who are more familiar with conventional materials and are simply unaware of the details surrounding the proper design of plastic parts (Fig. 1-8).

The role of the plastics processor in part design should be (minimally) in the area of review. The processor must be technically capable of suggesting appropriate design changes to create a producible part and to avoid the need for design changes after the production cycle has begun. Additionally, because the part design has a significant effect on the mold design, the plastics processor may be best qualified to highlight part-to-mold design issues prior to the design and machining of expensive tooling.

The part print should include proper tolerances (geometric tolerancing is now standard within the industry) and "critical" dimensions. Critical dimensions are those specified by the part designer to be highlighted on the print and inspected using approved statistical process control (SPC) methods (Fig. 1-9).

Another reason for the plastics processor to be involved in the design phase is to verify that the part design is suitable for manufacturing and assembly, if

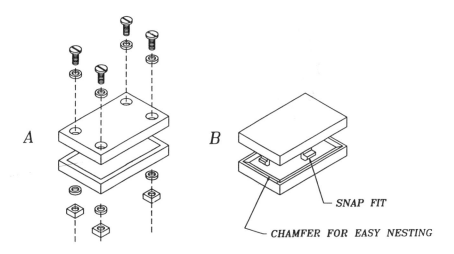

Fig. 1-10 Products designed with and without mechanical fasteners

required. In many instances, the plastics processor is directly involved with the assembly of several parts (all-plastic or plastic and metal). A part may be well designed if no further assembly is involved; however, it may be poorly designed to become a component of a larger assembly. Plastic part tolerances for molded products are different from those of machined metal. Plastic parts designed for assembly may include features to facilitate assembly, such as chamfers, integral hinges, snap fits, holes, and bosses (Fig. 1-10 and 1-11). These features should be designed into the part to reduce assembly time and to reduce the need for additional assembly parts, such as fasteners, clips, or screws.

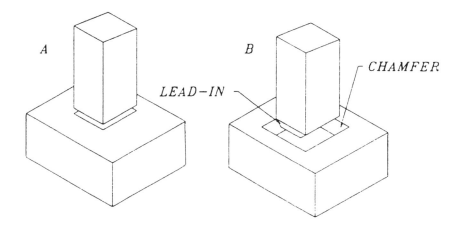

Fig. 1-11 Products designed with tapers and chamfers

2. Tooling Design Review

The tooling responsibility of the plastics processor varies as a function of its tool build capabilities. While the minimum involvement is to perform a tooling review, other aspects of tooling control may include tool design, tool build, and tool proofing or evaluation (the latter is discussed below).

In a tooling review, the plastics manufacturer carefully and thoroughly reviews all aspects of the tool design *prior* to the tool's manufacture. The tooling review has two main purposes: (a) to verify that the tooling will make the desired part with the quality specifications defined in the part print; and (b) to verify that the tool will fit the equipment and be as productive as required to manufacture the part within the cost structure defined. The latter consideration is often overlooked (completely or in part), both in new tool construction and in relocating existing tooling.

While experience is certainly important, a tooling checklist is critical. The use of a thorough checklist illustrates that the plastics manufacturer takes a proactive stance to avoid potential problems such as:

- Tool does not fit the equipment.
- Tool requires high maintenance.
- Tool is unable to control temperatures properly.
- Part production is inhibited.

3. Tool Proofing or Evaluation

The tool proof is the act of verifying that a tool is capable of producing plastic parts according to the part design specifications. In the past, a tool proof (e.g., for an injection mold) was the responsibility of the plastics manufacturer. Regardless of where the mold was built, the plastics manufacturer would mold several shots (after a stable molding cycle was established) and have all the part dimensions measured on all the cavities within the mold. Any discrepancies would either be corrected or be waived by changing the part print dimensions.

The tool verification procedures currently accepted within the plastics industry require that the tool builder be responsible for the tool proof. Additionally, the tool proof should use acceptable SPC techniques to conduct a machine capability study on the critical dimensions specified on the part print. All dimensions on all cavities (for an injection mold) should be measured at least once.

The tool proof requirement is normally specified in the purchase order contract. Any significant changes to the tool (due to repairs), the part print (due to print revisions), or the process require that a new tool proof or an abbreviated tool proof be conducted.

4. Product Costing

The basic cost structure of a plastic part is abbreviated MLOY:

- *Materials cost per part:* This could simply be the part weight multiplied by the materials cost. The part weight should include any nonproductive material components (e.g., sprue and runner).
- *Labor cost per part:* The labor assigned to one part is often derived by multiplying the direct labor rate assigned to a press (one operator may be able to operate more than one piece of processing equipment) by the hours required to produce one part.
- *Overhead cost per part:* The overhead for a plastic part might include allocated indirect labor (e.g., set-up/ repair and maintenance, quality inspection, and supervision), employee benefits, facility costs, capital depreciation, and supplies.
- *Yield loss per part:* Process inefficiency that results in scrapped parts is allocated to the yield category.

The costing of a plastic product can be complicated by many factors, such as the number of active cavities, the regrind allowance, and the types of secondary operations required.

Active Cavities. For most processes, the plastic part cost is developed based on a "full shot." If one or more cavities are unable to function or to produce quality parts, the molder must decide whether to stop the process and correct it, or allow it to continue. If the decision is to continue, the part cost can rise significantly, because there is less product per cycle and there may be additional nonproductive material. If the molder makes this decision, the price to the customer should not change. If the customer is asked to make the decision, the price of the part may increase significantly.

Regrind Allowance. Regrind is granulated thermoplastic material that has been previously processed. Nonproductive plastic materials (e.g., sprues, runners, and rejected parts) are reground using a granulator that usually resides next to the molding process. The granulator chops the material into flakes that can then be mixed with virgin plastic pellets and be reprocessed.

The amount of regrind allowed in the manufacturing of a plastic part should be specified on the part print. If no regrind is allowed, "No Regrind Allowed" should be noted on the part print. The addition of regrind to "virgin" plastic may lower the various properties of the plastic part beyond acceptable limits (Fig. 1-12). The effect of regrind on the cost of a plastic part should be understood and clearly stated by the manufacturer. In most cases, a manufacturer "builds" the cost of a plastic part by using enough regrind to consume all the nonproductive material in the process. Regrind is discussed in detail in Chapter 2.

Secondary operations include decorating, assembly, and packaging. The cost structure of a secondary operation is similar to that of the primary processing operation, except that it is calculated as the total cost (MLOY) of the plastic part plus any additional materials added to the process (e.g., paint, foil, or fasteners). The cost associated with a yield loss in a secondary operation can be extremely high, because in many cases it is impossible or impractical to recover the material in a plastic part that has already been decorated or assembled.

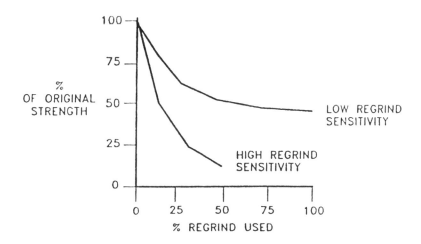

Fig. 1-12 Example of how regrind may affect the strength of a plastic part

5. Processing

Plastics processing can be grouped into the following core processes:

- Compression molding
- Transfer molding
- Extrusion
- Blow molding
- Injection molding
- Thermoforming
- Composite processing

These processes will be discussed in more detail in later chapters.

6. Product Assembly and Decorating

Secondary operations often are the least understood processes in a plastics manufacturing operation and result in the greatest yield loss. The assembly and decorating of plastic products are high-value-added operations that may significantly affect profitability (positively or negatively). Assembly includes welding, fastening, snap fitting, and machining, and decorating includes plating, printing/ coding, hot stamping/ adding decals, and painting.

7. Product Packaging and Shipping

All the technical expertise and time expended during product design, mold design, mold build, and molding of quality plastic parts can be lost by inappropriately packaging or shipping the product.

Type of Packaging. A plastic product may either be bulk packed or packaged directly after processing. This decision is based on the product design, the plastic material being processed, and the subsequent operations that will take place with the product. Bulk packaging of plastic parts is usually adequate for smaller parts that will cool quickly during the process and will not be damaged during impact and tumbling. In most cases, bulk-packaged plastic parts are shipped in the same containers used at the molding press. Plastic products that cannot or should not be bulk packaged must be packaged carefully in standard containers or, in some cases, in unique containers especially designed for one product.

Warping and Distortion. Unlike natural or conventional materials, plastics have a tendency to hold heat and to exhibit creep (deformation over time under a constant load at a constant temperature). Although a plastic part may appear to be cool and ready for packaging directly after processing, heat may be retained within the part walls, depending on the part design and the materials selected. Random handling and/or packaging after molding could cause the part to warp, or it could distort important features, rendering the product useless. Creep may result in part distortion simply due to the weight of the plastic parts within a packaging system.

Thermal Expansion. Plastics have a relatively high coefficient of thermal expansion. It is common for plastics designers to engineer a product to meet the environmental conditions of the end-use application, but to ignore the thermal extremes that may be experienced during shipping and storage. Semitrailers can easily experience temperature extremes from −40 to 180 °F. This temperature range could distort product and even cause loosening or failure of assemblies designed with dissimilar materials (e.g., plastic with metal fasteners).

Cyclic fatigue is product failure caused by repeated loading and unloading. It should be considered when plastic products must be shipped long distances, because the products may fail under the long-term, low-level cyclical loading conditions in a truck. It has been reported that plastic part design features, such as molded tabs and snaps, have failed during shipping excursions less than 1000 miles.

8. Materials Selection and/or Procurement

Choice of Supplier. Plastic molding compounds are obtained from five basic categories of suppliers:

- Plastic material manufacturers
- Plastic resin distributors
- Plastics compounders
- Plastics brokers
- Plastics reclaimers

Plastic material manufacturers are the facilities that actually manufacture the plastic compounds (e.g., Dow, DuPont, General Electric, Monsanto, Phillips, etc.). These manufacturers sell in large quantities and offer complete sales and technical services to their customers.

Plastic resin distributors are usually wholly-owned subsidiaries of the larger plastics manufacturers. The nature of the plastics business makes it difficult for small and midsize customers to receive the same level of attention as larger customers, so some manufacturers created distributor networks (e.g., General Electric created Polymerland). Through these distributors, the small customer can order plastic resin in quantities as low as 50 lb, yet call for technical assistance as required. Distributors often offer a limited portfolio of plastic materials, focusing on materials manufactured by the parent company and plastics provided by other manufacturers that do not compete with the parent company's products. The sales and technical service offered by most distributors is excellent.

Plastics compounders purchase large quantities of plastic resin from the plastics manufacturers, then customize or engineer it to meet special customer requirements. Typical customizing options for plastics include:

- Adding fiberglass and other reinforcements
- Adding special additives, such as flame retardants
- Adding special processing aids to facilitate molding
- Adding or creating unique color grades

In most cases the market for customized plastic material is not large enough to attract the plastics manufacturers. The sales and technical support of plastics compounders is usually excellent. Typical plastics compounders are LNP, A. Schulman, and Thermofil.

Plastics brokers neither manufacture nor compound plastic resin. They purchase "pockets" of material, such as excess inventories and odd lots of plastic, from any available source and sell them below retail price. The material available from a broker may be from several different (or even unknown) manufacturers. There is little or no technical support available, and batch-to-batch quality problems may arise.

Plastics reclaimers are a growing segment of the plastics supply business. Reclaimers purchase both plastic resin and plastic product (usually in the form of identified regrind), then mix and re-pelletize the material. They often add reinforcements, to improve lost mechanical properties, and colorants, to create visual uniformity. The reclaimed material may or may not have been tested and had its specific properties evaluated.

The use of reclaimed plastic may be the best choice for nonstructural or noncosmetic products. Lack of technical support and lack of control over the source and quality of the plastic make reclaimed plastic of limited value to most molders and manufacturers of top-quality structural plastic parts.

Choice of Container. Plastic materials are usually shipped to processors in one of five types of containers:

- 50 lb sacks (25 kg boxes)

- 300 lb drums
- 1000 lb gaylords (40 gaylords to a semitrailer)
- Bulk trailer trucks
- Bulk rail cars

The choice of container should reflect the plastics usage pattern and procedures for inventory control. Processors that use large volumes of plastic resin

Fig. 1-13 Melt indexer equipment. Courtesy of U.S.I. Technical Service Laboratories

(over 100,000 lb/ month) often elect to receive the material in bulk shipments and store the resin in silos. The storage and handling of smaller quantities (1000 lb and less) could lead to using the incorrect resin color and grade. Numerous smaller containers also require more warehouse space and more materials handling. To reduce the amount of different-colored resins that would otherwise have to be inventoried, the resin color is usually "natural," and the processor performs coloring operations.

9. Material Quality Verification

Quality verification of incoming materials should begin with the supplier. Certificates of compliance (C of C) should be requested by the processor for all incoming plastics. Most automotive quality plans demand that resin C of C be obtained from the supplier and maintained by the processor. The contents of the C of C should be agreed on by both the supplier and the processor.

Preshipment sample certification is another technique processors use to verify the quality of incoming plastic material. The processor and supplier jointly decide what tests the supplier will perform on a batch of plastic. Prior to shipping large quantities of a new batch of plastic, the supplier sends the processor a small sample of the resin (50 to 100 lb), along with the test results. The processor quickly examines the test results and molds and/ or tests the preshipment sample to verify its quality. The processor then allows the supplier to release the shipment of resin, according to an agreed-on shipping schedule, or the processor rejects the potential shipment.

Unless there is a special requirement, most processors do not perform extensive tests on the incoming plastic material. The one test that is quick and provides basic processing information is the melt index test, which measures the flow rate of plastic material under a constant load and temperature. The melt index equipment, which can range in price from $3,000 to $20,000, determines the basic flow rates of the material under specified conditions. The absolute value of the melt index is not as important as the relative values of the same material over different batches. Basically stated, the melt index tells the processor whether one shipment is different from prior shipments received.

Quality verification is also performed for material in process and for outgoing product. A basic concern of plastics processors is to avoid batch-to-batch variations in the plastic material. One of the tests that can be performed by the plastics processor (and that should also be performed by the plastic material supplier) is the melt index or extrusion plastometer test (ASTM D 1238). The equipment for this test is low in cost and simple to use (Fig. 1-13). Plastic pellets are melted and extruded under a set of controlled conditions.

The extrudate is created with a fixed orifice or die, and the processor weighs the extrudate over a specified length of time. If different batches of the same material produce significantly different amounts of extrudate, the processor has a basis for discussion with the plastic material supplier.

10. Material Compounding and Customizing

As discussed above, plastics processors may purchase their plastic compounds from companies that specialize in customizing and compounding plastic materials. However, it is also common practice for the processors to perform some compounding and customizing themselves. The typical categories of customizing and compounding of plastics include coloring, adding "regrind" to virgin plastic, and adding processing aids.

Coloring of plastic materials by the processor is often done to reduce the inventory of resins and to lower cost. Coloring can be accomplished in either a batch process (coloring 50 to 1000 lb at one time) or an in-line process (coloring material as it is conveyed to the process).

The colorants used must be compatible with the plastic resin to be colored, and they must not have too significant an effect on the mechanical properties of the plastic product being manufactured. Colorants may be dry or liquid; liquid colorants are often used because they are dispersed more easily within the resin system. Regardless of the type of colorant, the equipment used to add it to the plastic must be sophisticated enough to add it in the correct proportion by weight (as opposed to volume), and to sufficiently and homogeneously disperse the colorant throughout the plastic.

Adding regrind to virgin plastic is commonly done by plastics processors because it is simple and can be the difference between profit and loss on some products. During the granulating process, several control factors are often overlooked that can result in loss of material, loss of product, and even damage to the processing equipment:

- *Contamination* can result from mixing different plastics, nonplastic material (e.g., paper or cardboard), or metal (from the granulator itself or from foreign objects).
- *Wide variation in granulate particle size* results from improper granulator set-up. This can interfere with mixing and reprocessing.
- *Fines* (small dustlike particles of regrind, created during the granulation process), tend to heat up and degrade quickly when they are reprocessed, because of their small mass. Larger plastic pellets can overheat and burn during melting, causing black specks in the plastic

part that are both visually and mechanically degrading. Fines can be minimized by proper granulator set-up, screening, and loading of the regrind.

• *Fluff* (a low-density, fibrous form of regrind) may occur as the plastic is "shaved" or "skived" by the granulator. Fluff changes the regrind density significantly and can also prevent proper mixing. Fluff, like fines, should be removed prior to mixing.

• *Regrind in the final mix* results from compounding regrind by volume. All regrind should be compounded with the virgin material by weight only, because the bulk density of the regrind will be different from that of the virgin pellets. Regrind in the final mix can cause variation in the plastic product performance.

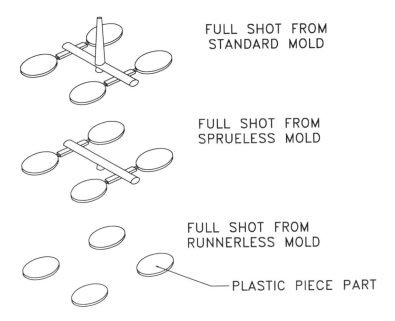

FULL SHOT FROM
STANDARD MOLD

FULL SHOT FROM
SPRUELESS MOLD

FULL SHOT FROM
RUNNERLESS MOLD

——PLASTIC PIECE PART

Fig. 1-14 Mold designs that reduce the amount of nonproductive material generated

Adding Processing Aids. These include:

- *Internal lubricants,* usually fatty acids (stearates) used to reduce internal friction and help control the melt temperature during processing
- *Antistatic agents,* used to lower the surface or volume resistivity of the plastic to prevent static charge build-up
- *Plasticizers,* added to plastics (usually vinyl) to increase the flexibility of the materials
- *Flame retardants,* added to plastics to enhance the flame resistance of the material (usually to meet regulatory agency requirements)

Processing aids and additives are discussed in more detail in Chapter 2.

11. Materials Conservation

Most plastics processes generate nonproductive material: sprues and runners, trim waste, culls, and rejected product. Many plastics manufacturers recognize the value of this material and address its re-use in a reactive manner, trying to reclaim it. However, the true cost/ energy/ material savings occurs when nonproductive material is not generated in the first place.

The elimination or reduction of nonproductive material often begins with the part design itself. The processor must be able to recognize how a part will be tooled and processed and attempt to reduce waste during the part design cycle. The processor must also consider how to eliminate or reduce nonproductive material during the mold design/ build phase of production (Fig. 1-14). Runners can be reduced in number or size, or even eliminated completely, by careful mold design/ build actions. Even older molds and dies may be candidates for upgrades to reduce nonproductive material. The common design measures taken to reduce nonproductive plastic material include:

- Reducing runner size (cross section and length)
- Replacing the sprue with either an extended nozzle or a hot-sprue bushing
- Replacing the runner and sprue system with a runnerless mold
- Using computer flow analysis to balance mold fill and reduce waste

Materials conservation efforts within a plastics manufacturing operation may not be recognized as a priority or as having cost-savings potential. The greatest deterrent is the perceived cost. The design and tool-building phases do cost more; however, the higher cost is more than offset by the long-term materials savings over the processing life of the product.

12. Equipment Selection and Procurement

Plastics processing is capital-intensive. Selection and procurement of the primary and secondary processing equipment is key to the success of the operation.

The complexities of plastics processing demand that the process be controlled by sophisticated microprocessor-based technology. The problem is that many manufacturers still own the low-tech relay-based machines and may not be able to justify buying newer equipment, as their current equipment may still have several productive years of service left. This dilemma requires one of these decisions:

- *Do nothing:* Remain with the older technology and risk losing customers and market share as quality requirements become more demanding.
- *Upgrade:* Retrofit the relay-logic machines with more sophisticated microprocessor-logic controllers. The cost to do this is of the order of $30,000 to $50,000, depending on the type of controllers and whether the labor is subcontracted or performed in-house.
- *Buy newer equipment:* This requires significant capital investment and may be a short-term financial burden, although there will be a long-term financial gain. The "newer" equipment may be either totally new or previously owned (for which there is a strong market).

If the decision is to buy newer equipment, the plastics processor must understand that the plastics equipment market is very dynamic. There are equipment manufacturers entering and exiting the market weekly. The plastics equipment market is also a global market, with European, Canadian, and Asian manufacturers making inroads into what used to be a U.S.-dominated arena.

In selecting equipment (foreign or domestic), these factors should be considered:

- Does the equipment maintain a process? Is the process control repeatable?
- Can a process be easily established? Are controls easy to understand?
- Does the equipment have all the features required, currently and in the future, to minimize the need for modifications later?

- Are the components English or metric?
- Can the tooling used with the equipment be quickly changed?
- Is the equipment compatible with shop goals and existing processing equipment? Will existing equipment fit, such as hopper and dryers? Are spare parts available?
- Is the equipment easy to maintain? Can basic parts (hydraulic and electric components) inventoried for existing machines be used?
- Will the company train both maintenance and set-up personnel?
- Will the company provide immediate sales and service attention?
- What is the history of the company: Are customers satisfied?
- What is the future of the company: Will it continue to exist?

The last two considerations are especially important for small and midsize manufacturers, yet are often overlooked. Several U.S. machinery manufacturers have either exited all or part of the plastics machinery sector in the last ten years. While there are usually provisions made to service the existing customer base, or the product lines are picked up by other equipment manufacturers, there could be a negative impact on a plastics manufacturer with such equipment.

In many cases the foreign equipment manufacturers are well-based financially and have a good future. However, several "copy-cat" foreign equipment manufacturers are producing low-cost plastics processing machinery whose productive life will not outlast the equipment's depreciation schedule.

13. Equipment Set-Up

The ability to establish a process on plastics processing equipment requires a combination of experience and knowledge of the product, the tooling, the material, and the processing equipment. Plastics processes, regardless of the processing equipment, may be categorized as follows:

- *New process* (based on new tooling)
- *Process improvement* (to improve quality and/ or productivity)
- *Adaptive processing* (process variation to accommodate some change in tooling or material)
- *Processing for different equipment* (tooling is moved from one machine to another)

The plastics processor should have a methodology for establishing a process. The type of machine process controls may affect the methodology, but sophisticated controls do not mean there is a sophisticated processing methodology. There should be ample documentation defining the values for key parameters, such as pressures, times, temperatures, speeds, and strokes. The documentation should also include the parameters for all ancillary equipment, such as dryers, temperature controllers, take-off equipment, and so on. All process documentation should be signed by a process engineer, and any cycle changes or output changes should be noted by the shop comptroller.

Set-up and changeover of plastics processing equipment should be accomplished in less than one hour, regardless of the product, tooling, or equipment size. Long set-up times indicate that the methodology is inadequate, based on industry standards.

14. Equipment Maintenance

Equipment maintenance procedures should be well documented, with an emphasis on proactive (preventive) maintenance procedures. Ideally each component of processing equipment should have a time meter attached, and maintenance activities should be performed by the clock. Maintenance should be logged and tracked, to highlight any chronic problems and to avoid major breakdowns.

Equipment operators and technicians should be allowed to take an active part in the care and maintenance of the equipment. Simple oiling and cleaning by the operator will create an involvement that will be supportive of good maintenance.

An inventory of expendable tooling and parts should be maintained, and parts ordered should be received within a day. The equipment and the shop should be clean, and there should be little or no hydraulic fluid or other liquids around the machine. Knobs, lights, and other control adjustments should be in place and labeled clearly.

All plastics processing machine manufacturers hold periodic classes for technicians. If you buy a new piece of equipment, the class is usually free.

15. Production Control

Production control of a plastics processing operation can either support productivity or tie production into knots. Scheduling equipment to meet customer delivery needs and maintain productivity is complex. The larger the product portfolio, the more complex the scheduling requirements. Many smaller shops buy or produce a basic software package to help manipulate equipment, tooling, material, and time to optimize the use of the equipment.

The just-in-time thrusts started by plastics processors have all but eliminated the long production run. To facilitate scheduling, most companies hold a daily production meeting that includes personnel from sales, production control, manufacturing, and engineering. This allows all areas to have input so scheduling decisions are timely and intelligent. The ability to efficiently schedule production of plastic parts often becomes a point of differentiation when potential customers assess plastics manufacturers.

2

Plastic Materials

Plastics are *polymers.* A polymer is an organic macromolecule (a large molecule based on carbon) comprised of several hundred or several thousand repeating segments, called mers, linked together in a chainlike form (Fig. 2-1). The number of times a particular segment repeats is referred to as n, the degree of polymerization. (Polymerization is discussed in the next section of this chapter.) As n becomes larger, the polymer molecule becomes longer and the molecular weight of the polymer increases. The chainlike molecule is often referred to as the carbon chain (Fig. 2-2). Compared to the molecules in common materials such as water or oil, the molecules that define a plastic are long ($n = 100$ to >1000).

To understand the effect of the length of plastic molecules, it is helpful to think of these long polymer chains as spaghetti cooking in a pot of boiling water. If the average length of each piece of pasta was 2 in., the cook would be hard-pressed to pick up very much on a fork. On the other hand, if the average length of each spaghetti strand was 8 in., the pasta would easily become entwined and entangled, allowing a larger amount to be gathered on the fork. The longer the polymer (plastic) molecules, the more they become entangled. The degree of entanglement helps give plastic its mechanical properties and the flow characteristics, such as melt viscosity, that define processing parameters.

Polymers are:

▼ Organic

▼ Large Molecules

▼ Chain-Like

▼ Made From "Building Block" Molecules

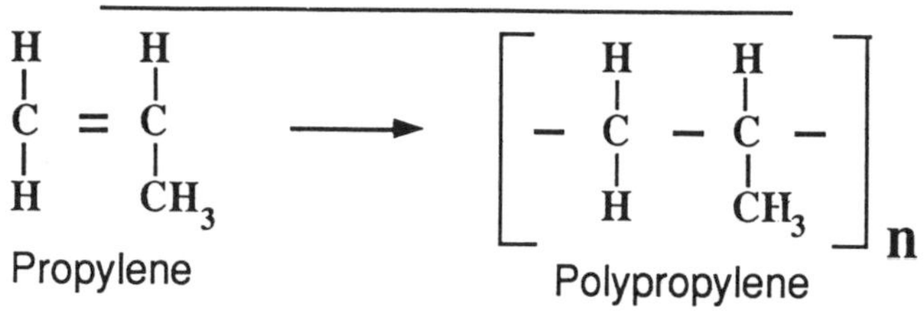

$$
\begin{array}{cc}
H & H \\
| & | \\
C & = & C \\
| & | \\
H & CH_3
\end{array}
\quad\longrightarrow\quad
\left[
\begin{array}{cc}
H & H \\
| & | \\
-\;C\;-\;C\;- \\
| & | \\
H & CH_3
\end{array}
\right]_n
$$

Propylene   Polypropylene

Fig. 2-1 Facts to remember about polymers

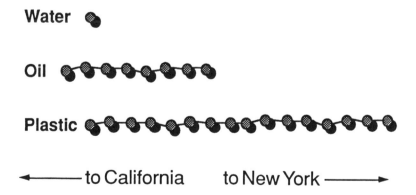

Water

Oil

Plastic

◄─────── to California   to New York ───────►

Fig. 2-2 The length of the carbon chain in plastic compared to that in water and oil

Each plastic is unique, and there are thousands of them. Understanding how plastics work will help the plastics processor select the best plastic for a particular application and establish the best processing parameters.

Feedstock Materials: The Source of Plastics

The feedstock materials for plastics fall into three basic categories (Fig. 2-3):

- *Petroleum,* usually in the form of a light distillate such as benzene
- *Natural gas,* in the form of methane
- *Agricultural materials,* in the form of wood or cotton (cellulose) or soybean byproducts

Petroleum-based feedstock prices, and the price of plastic materials, should be carefully monitored by large-volume plastics users (Fig. 2-4). Political factors, environmental concerns, and natural disasters have a major impact on feedstock availability, and therefore on pricing.

Feedstock materials are chemically developed into a monomer (single unit). The monomer is then reacted with a catalyst, heat, and pressure to create

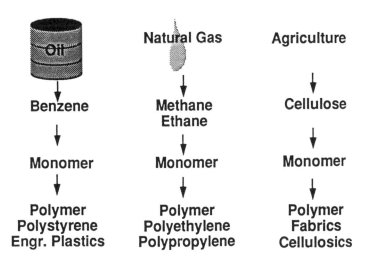

Fig. 2-3 Feedstock materials for plastics

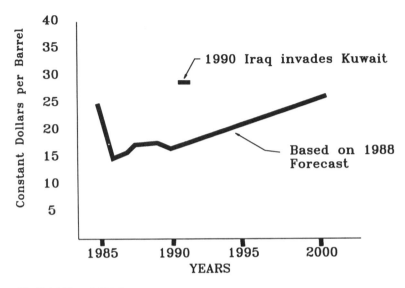

Fig. 2-4 Effect of OPEC oil pricing on the price of plastic

the polymer. This process is called polymerization. It is usually a batch-type process in which several thousands of pounds of polymer (plastic) are manufactured in large reactors. The plastic mass has to be granulated and then pelletized to a cylindrical shape with a diameter of approximately 0.060 in. and approximately 0.180 in. long. Prior to the pelletizing process, the plastic may be compounded with various additives, which are described later in this chapter.

Some plastics, such as polystyrene, can be polymerized directly into a spherical shape (see Fig. 1-1, in the previous chapter). These shapes allow the plastics processor to conveniently transport and handle the plastics in the various processing equipment. Also illustrated in Fig. 1-1 are the three common packaging systems in which plastic pellets are shipped: 50 lb sacks, 300 lb cardboard drums, and 1000 lb gaylords.

Materials Classification

Plastic materials are classified and subclassified to the point that even many plastic product designers become overwhelmed by the variety of chemical and trade names. To make sense of all the nomenclature, it is best to start with the two major plastics categories: thermosetting materials and thermoplastic materials.

Thermosetting Materials

Thermosetting plastics are plastic compounds that "set" or cross-link upon heating. The cross-linking process actually is the formation of chemical bonds between the long carbon chains (Fig. 2-5). The additional chemical bonds of the cross-linked thermosetting plastic allow it to absorb more thermal energy (heat) before the carbon chain is broken. For this reason, thermosetting plastics usually are able to perform at higher temperatures, which affords the plastic part designer a material with outstanding chemical and electrical resistance.

The cross-linking process is irreversible. Once set, the thermosetting plastic cannot revert to its prior stage. An analogy is the baking of a cake. Once the ingredients are mixed and the batter is baked, there is no way to fix or undo the process if the baker is not satisfied with the cake's quality. Terms often used to refer to the cross-linking process are *set, cure, vulcanize,* and *kick over.*

Typical plastics that are thermoset are listed in Fig. 2-6, with reference to the projected growth in their use.

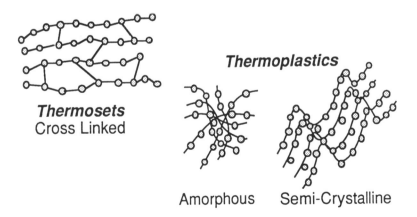

Fig. 2-5 Comparison of molecule structure in thermosets and thermoplastics

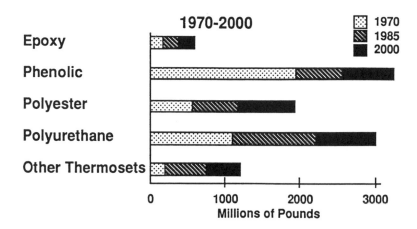

Fig. 2-6 Cumulative domestic thermoset growth. Source: "Plastics A.D. 2000," Society of the Plastics Industry

Thermoplastic Materials

Thermoplastics are plastic materials that soften upon heating and harden upon cooling. This process is reversible, just as ice hardens upon cooling but can be remolded by heating.

Unlike thermosets, thermoplastics have no chemical bonds between their long chain molecules (Fig. 2-5). However, the way the thermoplastic molecules position themselves next to each other, and the intermolecular forces that hold them together, do affect their properties and classification. Figure 2-7 highlights a variety of thermoplastic materials, with reference to the projected growth in their use.

Forms of Thermoplastics. Random entanglement of the thermoplastic molecules (much like the long spaghetti on the end of a fork) is called amorphous structure (without a logical order). Amorphous thermoplastics can be clear, have uniform (isotropic) properties in all directions, and have a melting range versus a melting point. Examples are polystyrene, acrylics, polycarbonate, and polyvinyl chloride (PVC).

Thermoplastics in which the molecules have an order are called semicrystalline. However, unlike the crystal structure in salt or metals, the molecules line up next to one another only occasionally (Fig. 2-5). Semicrystalline thermoplastics are characterized by opacity, nonuniform (anisotropic) properties,

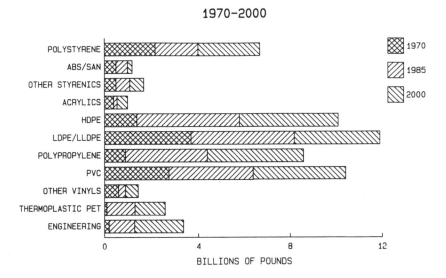

Fig. 2-7 Cumulative domestic thermoplastic growth. Source: "Plastics A.D. 2000," Society of the Plastics Industry

and distinct or narrow melting ranges. Typical semicrystalline thermoplastics are polyethylene, polypropylene, nylon, and thermoplastic polyesters.

Thermoplastic materials are found in a wide variety of other forms, including:

- *Elastomer:* able to stretch 2 times its original length and fully recover
- *Plastic:* the opposite of elastic (as in an elastomer); tends to remain in its new shape if distorted
- *Rigid:* stiff and maintains its shape
- *Flexible:* easily folded or distorted
- *Ductile:* can be stretched or pressed without losing its basic integrity

Using this information, one could describe polystyrene (the plastic used to make model airplanes) as a rigid, amorphous thermoplastic. Low-density polyethylene (the material used in plastic bags) could be described as a flexible, semicrystalline thermoplastic. A flexible plastic may not necessarily be an elastomer; however, virtually all elastomers are flexible.

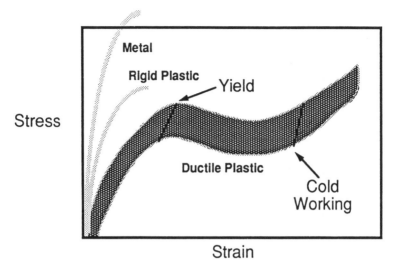

Fig. 2-8 Tensile properties of plastics versus those of metals

Figure 2-8 illustrates the difference in tensile properties between plastics and metal. Stress is the force per area applied in a tensile (pulling) mode, and strain is the resulting deformation (change in length) of any material.

Engineered Thermoplastic Materials. Most of the plastics used today are not pure polymers. To match the properties of a plastic to the requirements of a specific application, the polymer is often modified (custom engineered). There are two main techniques for enhancing a plastic: creating a copolymer and creating a polymer blend.

Creating a Copolymer. If a polymer is deficient in one or more of the properties required in an end-use application, it is often desirable to create a hybrid plastic by polymerizing two or more plastics together. A copolymer combines the best properties of each plastic into one material. An example of a copolymer is polyacrylonitrile butadiene styrene (ABS), which combines three polymers:

- *Acrylonitrile,* an acrylic polymer known for its clarity, colorability, and brittleness
- *Butadiene,* a rubber-like material that acts much like a shock absorber in ABS, thus improving impact resistance
- *Styrene,* a clear, brittle, low-cost polymer

Plastics Can Be Comprised Of:

Pure Polymer Base
(Polypropylene)

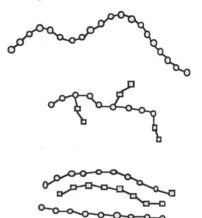

Copolymers
2 or more polymers
chemically joined
(ABS)

Polymer Blends (Alloy)
A physical Mixture of 2
or more polymers
(ABS + Polycarbonate)

Fig. 2-9 Two common techniques for custom engineering of polymers, copolymerization and blending

When polymerized together, these three different polymers are referred to as a terpolymer. They exhibit a phenomenon known as synergism, in which the abilities or properties of an entity are greater than the sum of its parts. In the case of ABS, the resulting physical properties (low cost, impact resistance, good colorability) are found in the product but not in all of the components.

Creating a Polymer Blend. Another technique for customizing plastics is to physically blend two or more polymers together. There is no chemical bonding in a blend, and a blend of polymers is often referred to as an alloy. A good example of a polymer blend is that of ABS and polycarbonate. The latter is known for its high impact resistance, natural flame retardancy—and somewhat higher cost. Customers needed a plastic that has better impact resistance and a better flammability rating than ABS, but the cost of polycarbonate could not be justified. It was determined that ABS and polycarbonate could be made compatible (not all plastics are compatible with each other) and could be physically blended to meet the requirements.

Figure 2-9 illustrates these two common techniques for custom engineering of polymers. In addition to creating copolymers and polymer blends, polymer engineers have been able to create graft polymers and branched polymers. Graft polymers are created by chemically bonding different materials at select points on the carbon chain in order to customize properties. Branched polymers have a modified carbon chain, with smaller chains or branches bristling off the main chain. These branches provide additional strength and rigidity to the polymer.

Additives and Modifiers

Even with all the efforts to engineer polymers to meet customer needs, there is still a demand for additional properties for specific applications (e.g., changes in strength, density, color, thermal properties, and cost). Therefore, most plastics used today are comprised of a polymer plus one or more additives or modifiers. Such combinations are known as plastic compounds. A plastic product designer must be familiar with the additives and modifiers, how they affect the polymer (positively and negatively), and how they will affect the plastic product. The major types of additives and modifiers are discussed below.

Reinforcements

Reinforcements are most often used to enhance mechanical properties, such as tensile strength and flexural modulus (rigidity), and thermal properties, such as deflection temperature. The reinforcement material is treated with a chemical coupling agent that helps it remain attached to the plastic matrix. The major advantages of adding reinforcements are improved strength, rigidity, and heat resistance. The major disadvantages are higher cost, shorter equipment and tooling life, and reduced product surface appearance.

Glass. The most popular reinforcement material for plastics is glass. It is used in the form of fibers that are compounded with the plastic, then pelletized for convenient use by plastics molders. Because the glass fiber is usually no longer than the plastic pellet (about 0.250 in.), the fiber can be molded with the same equipment and molds as those used for nonreinforced plastic. The two major glass fibers used in plastics today are E-glass, a low-cost fiber with good electric resistance, and S-glass, which is more expensive than E-glass but offers improved mechanical strength.

As with most other reinforcement materials, the addition of glass significantly erodes both the molding machinery and the mold. The long-term processing cost should be considered when working with heavily reinforced plastics. Another consideration for the plastics processor is the fact that intro-

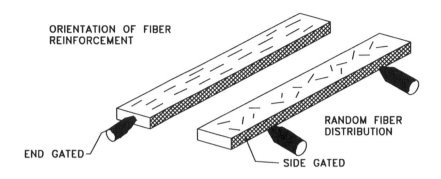

Fig. 2-10 Glass fiber reinforcement of plastic

ducing a fiber into the plastic melt stream may result in fiber alignment, which results in nonuniform physical properties (Fig. 2-10).

Carbon/Graphite. Used predominantly in advanced plastic materials, carbon/graphite fibers significantly improve a plastic's strength and rigidity.

Mica is a quartzlike mineral that is considered particulate in nature. A particle, unlike a fiber, is unlikely to orient during the molding process, and it therefore offers isotropic (uniform) properties.

Fillers

Fillers are materials added to plastics specifically to lower cost. They are compounded and pelletized with the plastic materials. Typical filler materials are wood flour, kaolin (clay), cotton, and cloth. The major advantages are lower materials cost, more product per pound of plastic (polymer), and improved heat resistance. The major disadvantages are lower mechanical properties than with unfilled plastic (depending on the filler used) and higher process variation.

Although fillers are more commonly used with thermosetting plastics, they are also available for thermoplastic materials. Sometimes a plastic compound contains a filler or filler(s) without the knowledge of the plastics processor.

Colorants

The science of coloring and color matching is complex. It often requires sophisticated equipment to ensure that colorants are both compatible with the material and are the correct type for a specific application. Coloring plastics is even more complex, because each plastic has a different color base. For example, acrylics, polystyrene, and polycarbonate are clear and therefore are readily colored. Nylon and ABS are naturally brownish in color and are more difficult to color. Thermosetting plastic and reinforced plastic are opaque, which limits the coloring of these materials.

The colorant itself is usually an organic or inorganic dye or pigment. The amount of colorant added to plastic material is relative to the color of the base plastic (i.e., the darker the base plastic, the more colorant required). Colorant is usually precompounded with the plastic, but if justified, it can be added to the plastic by the processor. The colorant may be in the form of a powder, liquid, or pellets. A typical colorant is added in the range of 1 to 4% by weight.

The major advantage of colorants is the ability to create products in a wide range of available colors and/ or with coloring throughout the material. Possible disadvantages are higher cost, difficulty in matching colors, batch-to-batch color variations, and change in colors when exposed to heat and sunlight.

Another disadvantage is that like all additives, colorants affect other properties. As an example, consider the colors white and black. White colorant is usually a form of titanium dioxide, and when it is added in sufficient loadings to effect a good white color, the plastic involved can become stiffer and suffer from a decrease in flexibility. Black colorant is usually a form of "carbon black," and it can increase the rigidity of many plastics. Although these secondary effects may be desirable, they must be understood. A product manufactured with a natural (uncolored) nylon plastic may perform differently than the same product manufactured with a white nylon material.

Flame Retardants

Plastic materials burn. Most designers want to be able to slow the burning rate in their selected plastic sufficiently to meet agency and consumer requirements. Flame retardants are added to plastics to slow the rate of burning and/ or to prevent the plastic from supporting a flame. The major advantages are reduced flammability and improved heat resistance. The major disadvantages are higher cost, possible processing problems, and reduced mechanical properties. Flame retardants fall into two main categories: compounds that when heated will generate a gas that starves a flame (i.e., removes the available oxygen); and plastics that are more flame-retardant than the selected plastic.

Flammability ratings for plastics are somewhat subjective and confusing. They are a function of wall thickness (e.g., a plastic with wall thickness of 0.125 in. may be rated more flammable than a plastic with a wall thickness of 0.0625 in.). The issues associated with the flammability of plastics are controversial, and the plastics processor must carefully consider the value of adding a flame retardant to a plastic versus selecting a more naturally flame-retardant material.

Stabilizers

Stabilizers are additives that help to control or enhance specific properties. As with all additives, there may be negative aspects to their inclusion.

Thermal stabilizers are used to improve the long-term stability of plastics when exposed to heat. The high-temperature resistance value of many plastics may be within a range 20 to 40 °F from the end-use temperature. The thermal stabilizer affords the plastic the widened temperature range required to meet an application. Thermal stabilizers are often used in heat-sensitive plastics such as PVC.

Plasticizers are chemicals added to a plastic (usually PVC) to make it flexible. Like many additives, plasticizers vary in their compatibility with the plastic compound. As a result, the plasticizer, over time, may become extracted from the plastic.

A good example of this phenomenon is automobile seats. When you purchase a new car, you immediately become aware of the "new car smell." This aroma is actually a combination of odors from carpeting, adhesives, paints, solvents, and the plasticizers used to keep the vinyl upholstery supple. (Even if you have leather seats, you still have vinyl in the padded dash!) With time, changing temperatures, and repeated sliding in and out of the car, the plasticizer is gradually removed from the upholstery. (The milky mist that lingers on the inside of the windshield of a new car on a hot summer day is plasticizer.) After enough plasticizer is extracted from the vinyl, the flexibility of the material decreases, and the seats and padded dash blister and crack. A new product, spray-on protectorates, has been developed to combat this problem. These are actually a form of plasticizer that coats the surface of the vinyl, preventing any further plasticizer from being extracted from the vinyl.

Ultraviolet (UV) Light Stabilizers. Many thermoplastics have a tendency to fade and/or physically degrade when exposed to sunlight. The UV stabilizer provides the plastic with the UV resistance it needs in order to be used in exterior applications. A good example of a plastic that requires UV stabilization is polycarbonate. This strong thermoplastic would not be able to compete against other plastic materials in applications such as auto taillight lenses if it were not for UV stabilization. Some colorants, such as carbon black, provide UV stabilization as a secondary benefit.

Antistatic Agents

One of the advantages of plastics is their inherent dielectric ability: they do not conduct electricity. However, this fact becomes a problem when static electricity needs to be dissipated. Electric charges that accumulate on the surface of a plastic will remain there until neutralized. Antistatic agents fall into three major categories: internal, external, and ion discharge (Fig. 2-11).

Internal antistatic agents are chemicals that are compounded in the plastic and migrate or "bloom" to the plastic surface due to their incompatibility with the plastic. As a result, the surface resistivity of the plastic decreases (i.e., the surface becomes more conductive). This change in the electrical characteristics of the surface is enough to dissipate the static charge. The internal antistatic agent has a finite life, and the effect gradually disappears. Internal antistatic agents have proven helpful in the recording industry, where static charges on

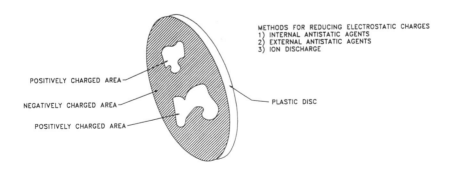

Fig. 2-11 Antistatic agents in plastic

the surfaces of phonograph records and compact discs increase the attraction of dirt and dust.

External antistatic agents are applied to the surface of a plastic part. The principle of decreasing surface resistivity is the same as with the internal antistatic agent. External antistatic agents are more effective, but they are also more short-lived. They have provided a niche market with the development of anti-cling fabric softener dryer sheets. These sheets temporarily increase the surface conductivity of polymer fabric, and the clothes lose their static charge.

Ion discharge antistatic agents developed for the electronics and packaging industries, which need short-term static charge dissipation. The ions are generated electrically or via low-level radiation. Plastic products are exposed to a slightly ionic atmosphere (such as an air stream) whose charge is opposite the charge of the surface electrons that cause static electricity. As a result, the static charge on the surface of the plastic is temporarily neutralized. This process is particularly useful in the dissipation of static charges on plastic parts with complex shapes.

Biocides

Many plastics attract undesirable lifeforms such as fungi and bacteria. Plastics have also been known to be food sources for undesirable creatures such as rodents. Manufacturers of underground conduits and plumbing products have found that biocides added to the plastic can provide adequate protection from these pests. A more common example of a biocide used in plastic is the insecticide used in pet flea collars. This biocide migrates to the surface of the extruded vinyl collar, providing the pet relief from fleas for several months.

Foaming Agents

Cellular plastic (foamed plastic) has been of particular value as a thermal insulation product, serving the building and construction industry and the packaging industry. All plastics can be foamed by the introduction of an additive or filler. There are three main types: internal agents, external agents, and microballoons.

Internal blowing agents are added to the plastic and decompose within a specific temperature range during molding. Their decomposition produces a gas (usually nitrogen) that forms a cellular structure when it is allowed to expand in the plastic melt.

External agents are usually gases (steam or nitrogen) that are physically introduced into the plastic melt to provide the cellular structure.

Microballoons or microspheres are small, hollow spheres (<0.010 in. diameter) that are mixed with the plastic (usually a thermosetting plastic) and become the cell structure for what is called a syntactic foam.

Regrind

As explained in Chapter 1, regrind is thermoplastic material that has been granulated and reintroduced into the process, usually by mixing it with virgin plastic. The introduction of regrind allows a plastics processor to get the maximum materials usage; however, the regrind is not exactly the same as the virgin material and may negatively affect the process. Compared to virgin plastic, regrind has:

- A longer history of heat exposure
- A different shape, so that it processes differently
- Reduced physical properties (Fig. 2-12)
- An altered color, or contaminants

The plastic product designer should specify the maximum regrind allowed for a particular part, directly on the print. This includes specifying "No Regrind Allowed" if that is what is required. Plastic part designers should consider the issue of regrind not only in the part design, but also in the process and tooling. Plastic parts can often be molded with little or no nonproductive material generated, so that regrind need not be considered.

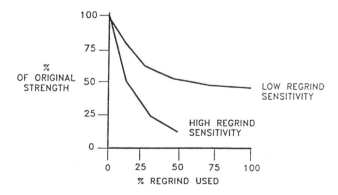

Fig. 2-12 Example of how regrind may affect the strength of a plastic part

The appendix to this chapter outlines a sample procedure for using regrind. It highlights some of the key issues and concerns encountered by plastics processors when using regrind material.

Appendix

Purpose. The purpose of this document is to outline a generic procedure for using regrind in a controlled manner to produce quality plastic parts.
Scope. This document discusses:

- Materials handling and identification
- Equipment control
- Mixing of virgin material and regrind
- Standard versus allowable regrind
- Product/ materials verification
- Reduction of nonproductive material
- Disposition of questionable material

Definition of Terms:

- *Productive material* is any and all material that is delivered to the customer as a product.
- *Nonproductive material* is any and all material that is not delivered to the customer as product, including sprues, runners, flash, pinch-off trim, and rejected parts.
- *Regrind material* is the granulated form of nonproductive material.
- *Virgin material* is plastic that has not yet been processed.
- A *generation* is created each time reground material is processed. First-generation material has been processed once, second-generation material has been processed twice, and so on.

Overview. It is the goal to reduce or eliminate nonproductive plastic material from being generated in a plastics manufacturing process. When nonproductive material cannot be eliminated, it is the goal to reuse this material by incorporating the nonproductive plastic back into the manufacturing process. This will be done within the constraints and limits established between the customer and the plastic part manufacturer. The reuse of nonproductive material shall not compromise the integrity of the plastic part. Any and all plastic parts produced using regrind shall meet or exceed the customer's specifications.

Materials Handling and Identification

Closed-Loop Handling. It is the goal to reduce the handling of nonproductive material by using a closed-loop materials flow system at the molding processes. In the closed-loop system, nonproductive material is granulated, mixed, loaded, dried (if required), and reprocessed alongside the plastics manufacturing process. The closed-loop system specifies that no material leaves the recycling process and that no nonproductive material is stored.

Open-Loop Handling. Plastic material that has been reground but cannot be put directly into the molding process requires a separate handling operation, called an open-loop handling process.

Labeling. The granulator and the plenum hopper should both be labeled with the following information:

- Material name (e.g., Zytel)
- Type (e.g., nylon 6/ 6)
- Grade (e.g., 30% glass-filled)
- Manufacturer (e.g., DuPont)
- Color (e.g., black)
- Part number (e.g., 123456-65)
- Amount of regrind allowable (e.g., 25% by weight)
- Initials of responsible personnel, with date

The labels should be printed on card stock in a standard format and mounted in a protective sheath. The labels should be located in plain view of the operator.

Warehousing and Inventory Control. Regrind that is generated in the production of plastic parts, but is not consumed in the production of those parts, may be packaged and stored for future use. The packaged regrind should be placed in a clean box that has been lined with a clean polyethylene liner that can be closed and sealed to prevent contamination. The label on the container should include the following information:

- Material name, type, grade, and color
- Date generated
- Date placed in storage
- Operator name
- Regrind part number

Regarding the last item on the list, regrind should never be stored, labeled, or inventoried with the same part number as that of the virgin material. A separate numbering system should be used for identifying stored regrind. One common method is to add the letter "R" as a suffix to the virgin plastic part number.

Regrind Generation. A regrind generation is created each time the material experiences a heat history (is processed). Zero-generation material is 100% virgin material, first-generation material has been processed one time, second-generation material has been processed two times, and so on.

Consider, for example, a plastic part design that allows for 50% regrind to be used. Assume that the part and process have been designed to have 50% productive and 50% nonproductive materials, and that the first time the part is molded it will use 100% virgin material. The regrind generations will develop as follows (numbers in parentheses indicate the generation level):

- *Zero generation:* 100% virgin material
- *First generation:* 50% virgin material, 50% regrind (1)
- *Second generation:* 50% virgin material, 25% regrind (1), 25% regrind (2)
- *Third generation:* 50% virgin material, 25% regrind (1), 12.5% regrind (2), 12.5% regrind (3)

Reusing Regrind. The greater the number of times a material is reprocessed (the greater the number of generations), the greater the risk that the properties of the plastic material will be reduced. This reduction will depend on the type of material, the type of process, and the type of product being molded. The properties that are typically reduced are impact strength, tensile strength, chemical resistance, and processability. Each material must be assessed to determine the maximum number of regrind generations that can be used in any specific application.

Avoiding Contamination. All boxes, barrels, gaylords, and other containers used to hold plastic material (virgin or regrind) should be covered at all times to prevent contamination from dust, dirt, oils, and other plastic materials. Clean canvas and plastic shrouds should be used to cover the plastic containers. Cardboard should be avoided, because cardboard itself is a common contaminant. The operator should be able to view the level of contents in the container and relocate any loader feeder tubes *without* removing the shroud.

Equipment Control

Cleaning granulators

- Before attempting to clean any granulator, the operator should be sure to unplug and/ or disconnect any electrical power. The guidelines of the granulator manufacturer should be followed and respected.

- Prior to disassembly of the granulator, the granulator material bin should be emptied and all regrind should be labeled and stored.

- Any plastic material removed from the granulator throat and cutting chamber should be properly thrown away, *not* reintroduced into the manufacturing process.

- Air hoses should not be used to clean the granulator. Blowing particles of plastic are unsafe and could contaminate other material. Instead, a shop vacuum cleaner should be used.

- After vacuuming, the granulator should be thoroughly wiped with a "tack cloth" to remove fines and dust, which could contaminate other material. A good tack cloth can be made by lightly spraying a clean cotton shop cloth with a commercially available antistatic spray.

Using granulators

- The plastic material that is placed into the granulator must be clean. Clean material is free of oil, grease, dirt, and particulate contamination, and it is not discolored or burned. Contaminated material must not be reintroduced into the process; it should be properly thrown away.

- A granulator is designed to cut plastic at a specified rate. Feeding it requires operator experience and knowledge. When placing material (usually in the form of sprues, runners, and parts) into the granulator, the operator must take care not to force the material down the throat of the granulator. If the material is forced into the cutting chamber, the granulator could jam or improperly cut the plastic.

- Granulator "stuffing tools," designed to aid the loading and moving of nonproductive material, should be used only if approved by the manufacturing engineer. The tools should be made in such a way and of such materials that they will not splinter or break. They must also not contaminate the material being granulated.

Granulator maintenance

- Granulator maintenance should be on a preventive basis to ensure both safety and proper cutting.
- The granulator cutting chamber should be inspected during every materials change and cleaning operation. The condition of the fixed and movable blades should be assessed and adjusted as required to meet the specifications of the granulator manufacturer. Blade sharpness, blade wear, and blade spacing must be monitored and adjusted as required. In addition, the screen should be inspected for wear and to ensure that it is the specified size, as measured by the hole diameter, for the process.
- All safety interlocks should be inspected every time the granulator is disassembled. No granulator should be used if the safety interlocks are not functioning properly.
- All sound-dampening shields should be inspected, well maintained, and properly replaced after the granulator is cleaned.

Loading systems

- All hopper loaders must be thoroughly cleaned during the materials change process or after every 48 hr of operation.
- Care should be taken to clean all air seals, filters, and loading lines. Loading lines that are not flexible or that cannot be readily removed should be cleaned by passing a "pig" through the line in a safe and proper manner.
- Operators should use a shop vacuum (not air pressure) to clean the hopper loader. Dust and fines should be removed with a tack cloth.

Plenum hoppers and cyclone separators

- The plenum hopper must be completely and thoroughly cleaned during each materials change. To be done properly, this procedure usually requires complete disassembly of the hopper.
- The internal screens and magnets must be removed and vacuumed to remove residual plastic.
- All hopper components may have to be wiped with an antistatic tack cloth to remove dust and fines.

- After reassembly of the hopper, the operator should carefully inspect the system to be sure that it is completely sealed. Improper assembly may result in leaks that could cause contamination or improper drying of the plastic material.
- Cyclone separators should be cleaned and maintained at the same frequency as plenum hoppers, following the same procedures.

Dryers

- Many plastics and elastomers are hygroscopic (they absorb water). This can interfere with processing and cause quality problems in the molded part. If the virgin plastic absorbs water, then the regrind of that material will also absorb water. The proper way to dry the material is by using a desiccant dryer attached to the plenum hopper.
- Dryer filters should be inspected during every materials change or after every 48 hr of use, whichever occurs first. Filters are used to trap dust and fines and are important in maintaining the efficient use of the dryer.
- Desiccant material, usually in the form of ceramic molecular sieves, should be on a preventive maintenance schedule. Inspection every six months and replacement as required is critical to proper drying.

Hopper magnets

- Hopper magnets must be used whenever regrind is introduced into the materials flow. As plastic is granulated, the metal cutting blades wear, and metal particles could be distributed throughout the granulate. Hopper magnets extract most iron-base metals.
- The magnets should be readily accessible and easily cleaned.
- The magnets may be located within the plenum hopper, or in a drawer magnet system located below the hopper and above the throat of the plastics processing machine.
- The magnets must be removed, inspected, and cleaned before every materials change or after every 24 hr of production, whichever occurs first.
- Care should be taken when cleaning magnets. The metal whiskers should be wiped off with an antistatic tack cloth, not blown, and be disposed of properly.

Mixing of Virgin Material and Regrind

All mixing of virgin material and regrind should be done by weight only. The reason is that the bulk density of regrind is less than the bulk density of virgin material. Bulk density, the density of plastic pellets before molding, is calculated by dividing the weight of the pellets by their volume. While virgin plastic pellets are uniform in size and shape (to allow for easy materials handling and hopper loading), regrind varies in form from flakes to large chunks. As a result, equal volumes of regrind material and virgin plastic pellets have different weights.

Mixing of virgin material to regrind must be accomplished by weight percentage only. If only time or volume is used as the proportioning variable, mixing will be incorrect and inconsistent.

Standard versus Allowable Regrind

Processing of Virgin Material/Regrind. The processing parameters used to manufacture plastic parts with 100% virgin material may have to be adjusted for the processing of various mixtures of regrind. A separate parameter set-up sheet may be in order when regrind mixtures are used. To minimize process and product variations, the same percentage of regrind should be used throughout the entire production run.

If there is insufficient regrind available to meet the maximum allowable regrind percentage, a lower percentage of regrind should be used. For example, even if 25% maximum regrind is allowed for a given production run, it may be preferable to consistently use only 15%. However, varying the percentage of regrind used may result in process excursions due to variation in melt flow.

Customer Part Prints. If regrind is to be used in a plastic part, the allowable regrind percentage should be specified by the part designer and noted on the part print (e.g., "25% Maximum Regrind Allowed"). If no regrind is to be allowed, the part designer should note this on the part print (e.g., "No Regrind Allowed" or "100% Virgin Material Only"). If there is no notation about regrind on a part print, it should be assumed that the part designer wants 100% virgin material.

Production Documentation. If regrind is used in the production of plastic parts, the percentage should be noted in the production records (e.g., "Manufactured using 25% regrind").

Manufacturer Specifications. Manufacturer specifications about the proportions of virgin material and regrind are only guidelines. Each plastic part and process must be assessed to determine the optimum proportion.

Product/Materials Verification

It is virtually impossible to verify that a molded part has been molded with regrind. Therefore, all controls for processing with regrind must be in place prior to molding. Any addition of regrind into a plastic part must be done only with the concurrence of the customer, and it should be so noted on the part print. Product testing, such as Gardener impact or falling dart impact testing, may indicate whether too much regrind and/ or too high a generation of regrind has been used. Such tests should be designed with the customer's assistance.

Reduction of Nonproductive Material

Reduction of nonproductive material is an ongoing process. Several steps can be taken to control the generation of regrind:

- Maintaining a clean work area to reduce the risk of contamination
- Proper cleaning of materials handling equipment during materials changeovers
- Proper maintenance of all equipment
- Covering all material containers to reduce the risk of contamination
- Proper training of all molding operators as to the proper handling and mixing of regrind

Disposal of Questionable Material

No questionable material should be reintroduced into the molding processes. It should immediately be removed from the processing area and properly thrown away.

3

Temperature, Pressure, and Time

Plastics processing, regardless of whether it is injection molding, blow molding, extrusion, or composite processing, requires the processor to be familiar with the key variables of temperature, pressure, and time. Most plastics processors use a set-up sheet that details the specific values for temperature, pressure, and time; however, many individuals involved in plastics processing do not fully understand these variables. This chapter will discuss the basic concepts that affect the use of these variables in the process.

Temperature

The first element in controlling temperature is to properly sense the true temperature. This is often difficult, because plastics processes are dynamic (always changing), and because heat is generated from two significantly different sources: the friction/ shear of the plastic being processed and the various heating elements used throughout the process.

Thermocouple wires are joined to form a junction which is electrically isolated from the sheath.

Ungrounded The junction is electrically isolated from the measured medium making it excellent for use in conductive solutions. Response time is slightly slower than for the grounded junction. Long life is achieved under conditions of rapid temperature cycling, shock or corrosion.

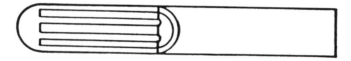

Grounded The thermocouple wires and sheath are welded together providing a rugged yet responsive sensing system. A good, general purpose junction for measuring liquids, solids or gases.

Thermocouple wires are joined to form a junction which is external to the sheath tip.

Exposed Fast response time is the feature of this junction. It is recommended for measurement of gas or solid surfaces in non-corrosive environments. Care must be used to avoid mechanical damage.

Fig. 3-1 Typical thermocouples. Courtesy of Watlow Electric Co.

Temperature Sensors

The three most common types of temperature sensors are the thermocouple, the resistance temperature detector, and the thermistor.

A thermocouple (Fig. 3-1) consists of two dissimilar metals joined at their ends. The metals are selected so that a measurable and predictable voltage can be generated at the junction. This voltage correlates to the actual temperature at the location of the thermocouple.

Thermocouples are the temperature-sensing devices most widely used in industry. The primary reasons for this are their:

- Low cost
- Ruggedness
- Ability to sense temperature in a wide range
- Dependability
- Fast response
- Simplicity
- Ability to be used with long extension wires

Drawbacks to the use of thermocouples include their:

- Weak ability to detect small temperature changes
- Instability with age
- Nonlinearity over wide spans
- Requirement that extension wires be the same wire as the thermocouple

In addition, a grounded junction or one exposed to grounding can create system problems if an electrical path (ground loop) forms between the sensor and the control circuit.

The performance of a thermocouple can be easily tested by placing the thermocouple sensor into boiling water. If the temperature reading is close to the boiling temperature of water (212 °F at sea level), it can be assumed that the thermocouple will be accurate at other temperatures. Another method of testing a thermocouple is to measure its millivolt output at a known temperature, using a millivolt-temperature conversion chart.

A resistance temperature detector (RTD) (Fig. 3-2) is a precision temperature-sensing device that is suitable for applications requiring accuracy, long-term electrical (resistance) stability, element linearity, and/ or repeatability. This wire-wound device displays a linear resistance change for a correspond-

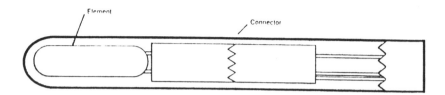

Fig. 3-2 An RTD. Courtesy of Watlow Electric Co.

ing temperature change. The coefficient of temperature is positive (that is, the resistance increases as the temperature increases).

An RTD is typically wound from platinum wire, which is inherently resistant to contamination and corrosion. A base resistance of 100 Ω at 32 °F is most commonly used; however, there are several characteristic curves of resistance change for temperature change. Selection of an RTD with an improper characteristic curve or improper base resistance can result in as much as a 20 °F error in the indicated temperature.

Drawbacks to use of RTDs include their:

- Slower response due to large element size
- High cost
- Sensitivity to shock and vibration
- Low sensitivity (small resistance change in response to temperature change)
- Low base resistance

Low sensitivity and low base resistance become a concern when long leads are required, because the leads create additional resistance. When added to the resistance of the RTD element, this can result in temperature indication errors. To overcome lead length problems, a three-wire RTD sensor should be used, so that the third wire senses lead resistance. When using two-wire RTDs in a three-wire system, S-2 and S-3 can be jumped together, but the resulting lead resistance can cause indication error. For each 0.3 Ω total lead resistance, an approximately 1 °F indication error will occur.

A thermistor (Fig. 3-3) is manufactured from a mixture of metal oxides, then encapsulated in epoxy or glass. The result is a sensing device that displays

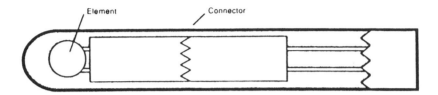

Fig. 3-3 Thermistor. Courtesy of Watlow Electric Co.

a very direct, nonlinear relationship between resistance and temperature. The coefficient of temperature is usually negative (the resistance decreases as the temperature increases).

Thermistors are small, become more stable with use, and exhibit great sensitivity to small temperature changes (the element is highly accurate). Base resistance values are usually much higher than those of RTDs, which allow long leads to be used without appreciable calibration error. These characteristics make the thermistor a very responsive sensor that permits control of a process variable to within 0.5 °F or better.

Properly matching a thermistor to an instrument is much more critical than for an RTD. Element resistance can vary from one thermistor to another by a factor of as much as 100 to 1. Improper selection of a thermistor element could result in an error of like magnitude.

Thermistors are generally not interchangeable, and unless additional instrument circuitry is added, a thermistor will not provide a fail-safe condition if the element should open. Other drawbacks to the use of thermistors are their fragile nature, limited temperature span, initial element drift, and decalibration at higher temperatures.

Thermal Control Principles

Control Devices. The selection of a temperature control device depends primarily on the degree of control required to achieve the desired results. Selecting more control than is actually required for the application will add needless expense and complexity to the system.

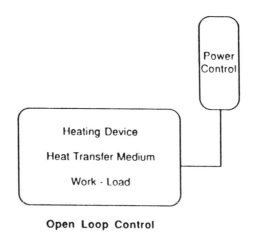

Open Loop Control

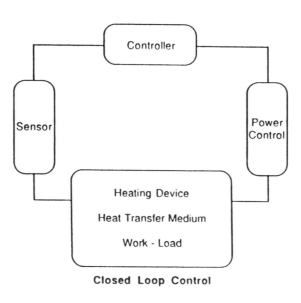

Closed Loop Control

Fig. 3-4 The two basic types of temperature controls

Temperature controls are of two basic types: open-loop and closed-loop (Fig. 3-4). A closed-loop device is self-correcting: as process temperature changes, a feedback loop provides up-to-date status information to the control component, which adjusts itself to regulate the system. An open-loop device has no self-correcting mechanism and is a much less desirable approach to temperature control.

Control Modes. The most common control modes are on/off and time proportioning.

On/Off Mode. As its name implies, the output device of an on/off control is either fully on or fully off. Temperature is always controlled about a setpoint (Fig. 3-5), dictated by the switching sensitivity of the on/off control. There will be a certain amount of temperature overshoot and undershoot, depending on the characteristics of the entire thermal system.

Temperature sensitivity (hysteresis) is designed into the control action between the on and off switching points, to prevent the output device from switching on and off within a temperature span that is too narrow. Switching

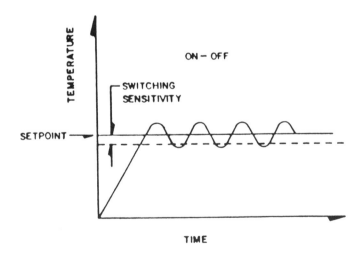

Fig. 3-5 Time vs. temperature profile developed by on/off control. Courtesy of Watlow Electric Co.

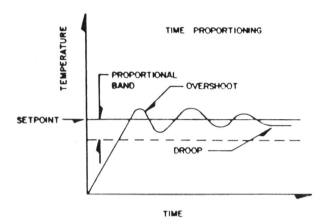

Fig. 3-6 Time vs. temperature profile developed by time proportioning control. Courtesy of Watlow Electric Co.

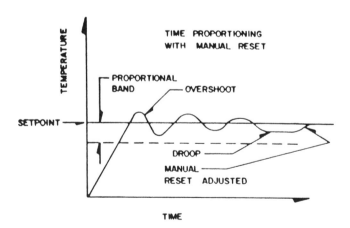

Fig. 3-7 Time vs. temperature profile developed by time proportioning control with manual reset. Courtesy of Watlow Electric Co.

repeatedly within such a narrow span creates a condition known as output "chattering" (intermittent rapid switching).

Time proportioning mode provides more precise control of process temperature. It operates in the same way as an on/off control (Fig. 3-6). When the process temperature approaches the setpoint and enters the proportional band, the output device is switched on and off at the established cycle time. At the lower limit of the band, the on time is considerably greater than the off time. As the process temperature more closely approaches the setpoint, the ratio of on/off time changes; the amount of on time decreases as the off time increases. This change in the effective power delivered to the workload provides a throttling-back effect that results in less process temperature overshoot.

The on/off cyclical action continues until a relationship of equal on/off time is achieved. At that time, the system will be stabilized such that process temperature is controlled at a point just below setpoint. The process temperature does stabilize with the resultant "temperature droop." This condition will remain, providing there are no workload changes in the system.

Reset: Compensating for Droop. If temperature droop cannot be tolerated, there are ways to compensate for it. Reset is a control action in which the rate of change of output is proportional to the input.

Manual reset must be performed by the operator. This adjustment brings the process temperature into coincidence with the setpoint (Fig. 3-7). Once the manual reset adjustment is made, it will ensure coincidence within a narrow span of the setpoint for which it was adjusted. If the setpoint is changed drastically (75 °F or more), coincidence between setpoint and process temperature will be lost, and the manual reset adjustment will have to be made again.

Automatic reset (integral) is performed by the control device itself, to compensate for a droop condition before it exists. An integration function automatically drives the process temperature up to setpoint, but automatic reset is prevented until the process temperature enters the proportional band (Fig. 3-8). If it was allowed to take place at any point in the span of control, a condition of extreme temperature overshoot would occur. The function of eliminating automatic reset is called *anti-reset*.

Rate: Compensating for Overshoot. As the process temperature graphs illustrate, temperature overshoot can occur with any control mentioned thus far. This condition may be hazardous to certain processes, so it cannot be tolerated. It is preventable with rate (derivative), a control action in which the output is proportional to the rate of the change of input. Rate measures the rate of increase of the process temperature and, if necessary, forces the control into time-proportioning mode in order to slow the increase (Fig. 3-9). This prevents

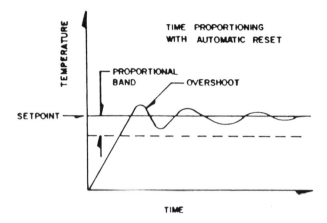

Fig. 3-8 Time vs. temperature profile developed by time proportioning control with automatic reset (integral). Note that the condition of "temperature droop" does not exist in this graph. Courtesy of Watlow Electric Co.

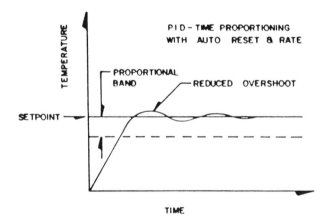

Fig. 3-9 Time vs. temperature profile developed by time proportioning control with automatic reset (integral) and rate (derivative). The effect of automatic reset is evident in the lack of "temperature droop," and the effect of rate is evident in the reduced amount of process temperature overshoot on start-up. Courtesy of Watlow Electric Co.

a large degree of overshoot on start-up and also prevents overshoot when system disturbances would tend to drive the process temperature up or down.

For applications that require precision temperature control or result in frequent system disturbances, select a time-proportioning temperature control with automatic reset and rate.

Heat Sources

Proper selection of a heat source is critical for process control, energy efficiency, and operator safety. It is prudent for plastics processors to assess their heater requirements on an annual basis to determine that they are properly matching their heater systems to their requirements.

Heater bands, in particular those supplied with new pieces of plastics molding equipment, may be adequate for the initial plastic parts and materials processed on the equipment. General-purpose heater bands, such as those supplied with plastics processing equipment, usually work accurately and efficiently for temperatures up to 600 °F. Many of the newer plastic materials require process temperatures over 650 °F, and in these cases special high-temperature bands may be required.

As energy costs rise and electricity becomes a larger component of overhead cost, many companies are considering replacing conventional heaters with high-efficiency heaters. The initial cost of high-efficiency heating elements is greater than the cost of general-purpose heaters, so the overall cost savings may take years to materialize.

Heater cartridges are common in molds and dies that require temperatures between 200 and 450 °F. As with a heater band, the heater cartridge should be carefully selected on the basis of temperature requirements, applications, and efficiency. It is critical that the heater cartridge have the proper fit within the mold or die. A slip fit is recommended for proper heat transfer. If the fit of the cartridge is too loose, there will be inefficient heat transfer, and hot spots could shorten the life of the cartridge. If the fit of the cartridge is too tight, the heater may be damaged during installation.

Water is the most common heat transfer medium for molds. It can be used because most molds require process temperatures between –20 and –200 °F. The water is cooled with a chiller unit, conditioned with either ethylene glycol or propylene glycol (antifreeze agents), and pumped through the mold. When temperatures above room temperature are required, water is heated by electrical immersion heaters and pumped to the mold (see Chapter 9).

Hot oil can be used when temperatures between 150 and 400 °F are required. The oil is heated via electrical immersion elements or natural gas burners, then pumped through the process. Hot oil presents several safety concerns, and processors should carefully assess whether electrical heaters would be more prudent.

Natural gas is a viable heating source for many plastics processes. It is most common in the thermoforming process, where plastic sheet is heated but not melted, but it can also be used in other plastics processes, such as injection molding or extrusion, and in ancillary operations, such as drying. Unfortunately, many equipment manufacturers have not developed or marketed gas-fired plastics processing equipment at a level to make it attractive to small and medium-size processors.

Microwave heating is under development for plastics processors, especially in the area of materials drying and curing, and it should be available by the year 2000. Microwave drying has the advantage of decreasing drying times from 3 hr to 30 min, which is a major improvement in efficiency.

Pressure

All plastics processing involves controlling pressure, whether it be die pressure, clamp pressure, cavity pressure, or system pressure. Unfortunately, pressure control is probably the variable that is least well understood.

Most plastics processing machines are hydraulic systems. The hydraulic oil, which is commonly at a pressure of 2000 psi, is used to turn screws, move barrels, or close clamp units. The 2000 psi hydraulic pressure is ported and valved to vary the pressure up and down; therefore, the system pressure is not necessarily the same as the processing pressure. For example, in many injection molding machines there is a 10-to-1 reduction in area from the rear to the front of the injection screw. Even though the pressure gauge on the machine indicates to the processor that the hydraulic pressure for injection is 1200 psi, this is not the injection pressure; it is the system pressure. The actual injection pressure is 12,000 psi, due to the 10-to-1 reduction in area at the tip of the screw compared to the diameter of the hydraulic cylinder.

The ability to control a process requires that the pressure first be accurately sensed and assessed. Historically, pressure has been measured with pressure gauges that monitor pressure in an analog fashion. Therefore it is very difficult, if not impossible, to determine the value of the pressure at any specific time in the process. The difficulty is compounded by the fact that any particular pressure is used for a very short time, sometimes less than one second. To overcome this problem, electronic pressure sensors have been developed that can have either a digital or analog output. Simply stated, the analog sensors measure pressure values that vary, to generate a pressure curve, while digital sensors measure specific pressure values in a manner that is directly compatible with computer systems (Fig. 3-10 and 3-11).

Fig. 3-10 An analog pressure sensor. Courtesy of CMC Technologies, Ltd.

Figures 3-12 and 3-13 illustrate pressure control systems. Pressure control is also discussed in the chapters about various processes.

Fig. 3-11 A digital pressure sensor. Courtesy of CMC Technologies, Ltd.

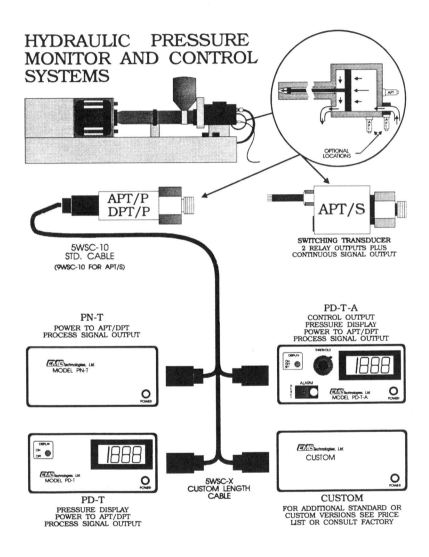

Fig. 3-12 An hydraulic pressure control system. Courtesy of CMC Technologies, Ltd.

IN-CAVITY MELT PRESSURE CONTROL/ MONITORING SYSTEMS

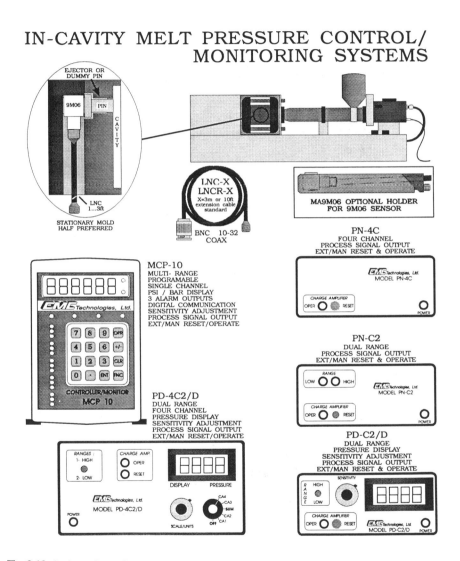

Fig. 3-13 An in-cavity melt pressure control system. Courtesy of CMC Technologies, Ltd.

Time

Time is critical in all plastics processes. Timers determine how long one aspect of the process has been in effect and when another aspect of the process should begin or end. Older plastics processes used mechanical timers, which were essentially clocks with a series of switches that timed process events. These timers were inaccurate and prone to mechanical failure, resulting in malfunction of the switching mechanisms and poor process control.

Today, digital timers can be retrofitted or initially incorporated into plastics processing equipment. Digital timers are both accurate and precise; most are capable of timing to 0.01 sec. The data are directly usable by computers/microprocessors. Additionally, the technology for switching from function to function is also now electronic, which means that there are no moving parts to wear out and that precise control can be maintained.

Fig. 3-14 A plastics processing control system that integrates temperature sensors, pressure sensors, and timers. Courtesy of BASF

Temperature sensors, pressure sensors, and timers can now be integrated in process control systems (Fig. 3-14), which results in higher-quality, more efficient plastics processes.

4

Extrusion

More plastic resin is processed through the extrusion process than through any other plastics manufacturing technique. This fact, however, is misleading. Extrusion is accountable for significant quantities of plastic products, such as plastic film, sheet, and profiles, but it is also used to produce the plastic pellets that are later used by all the other plastics manufacturing processes. Therefore, the majority of plastic material produced in the world today has had at least one excursion through an extruder, regardless of whether the resin will ultimately be used by injection molders, blow molders, or extrusion processors.

Extrusion is one of the oldest manufacturing processes. Like compression molding (e.g., pressing earth into bricks), it dates back well before the birth of Christ. Tubular products, such as crude clay pipes, and food products, such as spaghetti and macaroni, are examples of early extruded products. Even today we find similar items, including plastic pipes and potato chips, made with extrusion methods.

Early extruders were not much more than push rams used to force plastic resins through a heated barrel and die. These ram extruders offered very little in terms of process control or uniform flow of the plastic melt, but they did provide manufacturers with an alternative to metal extrusion. The rubber industry has also employed extruders for over a century, but that industry

Fig. 4-1 Features of a single-screw extruder. 1, Die or accessory temperature controls. 2, Ammeter for each heating zone. 3, Motor load ammeter. 4, Screw speed tachometer. 5, Instrument panel. 6, Premounted, prewired panel for single power drop installation. 7, Hard surface screws. 8, Swing gate with heater. 9, Bimetallic-lined cylinders. 10, Melt pressure gauge and rupture disc standard. 11, Cooling fans. 12, Temperature controllers. 13, Melt temperature indicator. 14, Hopper. 15, Helical gear reducer. 16, Water-cooled feed section. 17, Window heaters. 18, Motor fan. 19, Machine base. Courtesy of Welex, Inc.

often refers to extruders as tubers, based on their use in the production of rubber hose and tubes.

Today's extruders are sophisticated pieces of processing equipment that use microprocessor-based control systems to ensure consistency of the extruded product. Extruders range in size from tabletop experimental units to massive high-volume production machines that may be longer than 50 ft. Regardless of the size of the extruder, there are several common features, as depicted in Fig. 4-1.

The Extrusion Process

The extruder is a melt pump: plastic pellets are melted, and the melt is forced through one or more dies that produce the shape of the plastic product.

Virtually all of today's extruders employ the screw method of plastication, using one or more screws that are specifically designed for the plastic resin being used and the plastic product being manufactured. This chapter will focus only on the screw method.

The process of extrusion is best understood if it is broken into key steps:

1. Material enters the extruder through the throat, an opening that links the materials hopper to the extruder barrel.

2. Once inside the extruder barrel, the plastic pellets are conveyed, via the turning screw, to the front or die end of the extruder.

3. During the journey to the front of the extruder, the plastic pellets experience a significant amount of friction, caused by the shear action of the screw and barrel. This frictional energy develops sufficient heat to melt the plastic. At the midpoint of the screw, 50% of the pellets are melted. In the final third of the barrel, the screw rotation has completed the melting of the plastic pellets. The average single screw extruder develops 85 to 90% of the energy required to melt the plastic. The electric heater bands on the outside of the barrel serve only as thermal blankets and to facilitate start-ups.

4. Now melted, the plastic melt is pumped through a die to make the plastic product.

5. While the extruder is forcing plastic melt through the die, the extruded product is being pulled by ancillary takeoff equipment. This allows the plastics processor to control output rate as well as part dimensions. Increasing the pull rate of the takeoff so that it is above the rate of the extruder output causes the extrudate to neck inward, which is often desirable to compensate for die swell. (Die swell is discussed in the section "Tooling for Extrusion" later in this chapter.)

6. Located between the extrusion die and the takeoff equipment is a cooling station. For horizontally extruded product, the cooling station is often a tank of water. The water is allowed to flow to keep its temperature constant. If the extrudate is vertically produced, the cooling station will usually be ambient air.

Parts of an Extruder

The Screw

The development of the screw extruder allowed plastics and rubber processors to increase both their productivity and the quality of their extruded products. There is no technical comparison between the old plunger-based extrusion machines and the newer screw-based extruders.

Early in the 20th century, the rubber extruders (tubers) employed a simple screw design (Fig. 4-2) that had a constant channel depth, a constant root diameter, but a varying lead. These early rubber processing screws were designed to melt and pump the rubber materials. As thermoplastic materials were developed and produced in large enough quantities, the screws designed for rubber processes were adapted and used to accommodate the new plastic resins.

The early plastics screws (Fig. 4-3) usually had a constant square pitch (lead diameter) and no distinctive feed or metering section. The root depth gradually changed from the feed to the discharge end, which resulted in compression of the plastic material. This compression facilitated the melting process.

Screw Design. By the 1960s the thermoplastic extruder screw had evolved in design to include three distinct sections (Fig. 4-4):

- *A feed section* with a constant depth
- *A transition section* with a varying depth
- *A metering section,* again with a constant depth

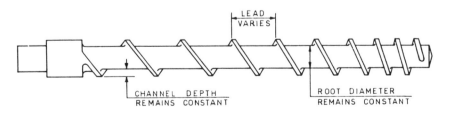

TYPICAL RUBBER SCREW

Fig. 4-2 Typical screw used in the extrusion of rubber. Courtesy of Spirex

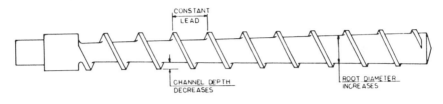

"PLASTIC" SCREW

Fig. 4-3 Early screw used in the extrusion of plastic. Courtesy of Spirex

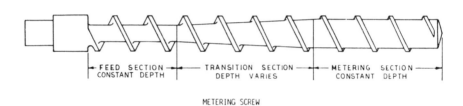

METERING SCREW

Fig. 4-4 Metering screw used in the 1960s. Courtesy of Spirex

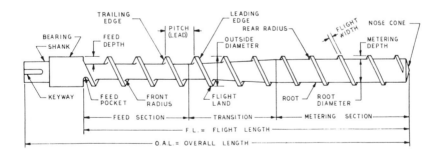

NOMENCLATURE: EXTRUSION SCREW

Fig. 4-5 General-purpose extrusion screw. Courtesy of Spirex

This screw design and its metering section helped to smooth out pressures, temperatures, and output. Variations of this screw design are the foundation for today's extrusion and injection molding screws. A typical, general-purpose extrusion screw design is depicted in Fig. 4-5.

Screw nomenclature is as follows:

- *Key way:* locates and secures the screw into the drive mechanism
- *Bearing shank:* precisely ground diameter that secures the screw radially
- *Feed pocket:* the rear of the feed section; defines the start of the flighted portion of the screw
- *Feed depth:* initial screw depth; used to calculate compression ratio
- *Front radius:* flight radius in the front of the flight
- *Pitch (lead):* distance between flights
- *Flight land:* width of the flight (flat section)
- *Leading edge:* front edge of flight land
- *Outside diameter:* the major diameter of the screw
- *Rear radius:* flight radius at the back of the flight
- *Root:* main shank or center shaft of the screw
- *Root diameter:* diameter of the screw root; usually varies in size
- *Flight width:* size of the flight land
- *Metering depth:* distance from the top of the flight to the screw root
- *Nose cone:* shape of the front of the screw
- *O.A.L.:* overall length of the screw
- *F.L.:* flighted length of the screw
- *Feed section:* first section of the screw; conveys plastic pellets
- *Transition section:* second section of the screw; conveys both plastic pellets and plastic melt
- *Metering section:* last section of the screw; conveys plastic melt
- *L/D ratio:* relative length of the screw (pronounced "the L over D ratio")

$$\text{LD ratio} = \frac{\text{Flighted length of the screw}}{\text{Outside diameter of the screw}}$$

- *Compression ratio:* helps define how much the screw will compress or squeeze the material; key variable for different plastics

$$\text{Compression ratio} = \frac{\text{Depth of feed section}}{\text{Depth of metering section}}$$

Screw Function. The screw of the extruder performs a number of functions: conveying plastic pellets, melting plastic pellets, conveying the plastic melt, and mixing the plastic melt.

Conveying Plastic Pellets. Conveying the plastic from the throat and feed zone occurs by the augur-like action of the turning screw. At first the cold plastic pellets are easily advanced within the barrel, but as the diameter of the screw root increases, they have a significantly restricted space in which to move. Combined with the die resistance at the front of the extruder, this retards the forward motion of the pellets. As a result, they remain in the barrel for a longer time and are subjected to the shear forces caused by the turning screw and the stationary barrel wall.

Melting Plastic Pellets. The shear forces result in friction between the plastic pellets, which produces sufficient heat to melt them (Fig. 4-6). The area between the screw flights in the transition section of the screw becomes a turbulent vortex of newly melted plastic, located just in front of the flight of the screw, and unmelted plastic, located in back of the flight of the screw. The interface of the plastic melt and the colder plastic pellets moves as the plastic

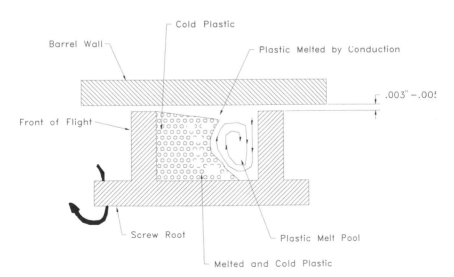

Fig. 4-6 How a screw contributes to the melting of plastic

is conveyed toward the front of the screw. By the time the plastic reaches the metering section of the screw, it has all been melted.

Conveying Plastic Melt. Approximately 80 to 90% of the heat required to melt the plastic pellets is generated from this frictional heat, not from the electric heater bands. The electric heater bands facilitate the start-up of the process, and they act as a thermal blanket to prevent or reduce loss of heat generated by the shearing action of the screw. The screw's ability to convey the plastic material decreases as the form of the plastic changes from cold pellets to hot plastic melt.

Mixing the Plastic Melt. The screw provides a significant amount of mixing, simply by the nature of its design and the melting of the plastic pellets. The ideal objective is to have the screw provide both thermal homogeneity, meaning the plastic melt has the same temperature throughout, and physical homogeneity, meaning the plastic melt is thoroughly mixed. To understand physical homogeneity, imagine that while an uncolored plastic material is being extruded in a stable process, a small amount of colorant is added. The color change of the extrudate shows how effectively the screw is mixing the material and how uniform the flow is through the die. Figure 4-7 illustrates three patterns of extrudate flow and material mixing.

Improving Mixing Effectiveness. Several techniques are available to improve the effectiveness of screw mixing. One of the most common is to alter the metering section of the screw to increase the mixing action.

The Dulmage screw (Fig. 4-8), developed by Fred Dulmage of Dow Chemical Co., was one of the first mixing screws. It has a series of semicircular grooves, cut into a long helix in the same direction as the screw flights. There are usually three or more sections, interrupted by short cylindrical sections. These interrupt the laminar flow and divide and recombine the plastic melt many times. In this way the screw works like a static mixer. It is still used in processing foam materials.

Mixing Pins (Fig. 4-9). In the early 1960s, several companies started to place radial pins in the screw root. These pins tend to interrupt the laminar flow and provide more effective mixing. They also make it possible to design the screw to work a little deeper and thus provide more output with the same degree of mixing. Many patterns and shapes of pins have been used, but in general, they are placed in rows around the screw. They are located in the metering section, so they contact the material after most of the melting process has occurred. A typical arrangement has three rows: one row at the beginning of the meter, another one flight back from the end of the screw, and another halfway

Extrusion Flow Patterns

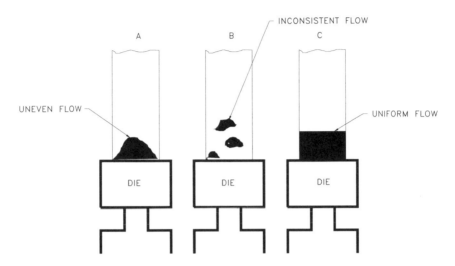

Fig. 4-7 Three patterns of extrudate flow and material mixing. (a) Good material mixing, but uneven flow rate across the die. (b) Uniform flow rate across the die, but poor mixing. (c) Uniform flow rate and good mixing

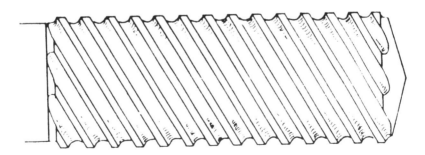

Fig. 4-8 Dulmage screw. Courtesy of Spirex

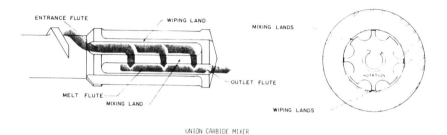

ENTRANCE FLUTE   WIPING LAND   MIXING LANDS

MELT FLUTE
MIXING LAND   OUTLET FLUTE   WIPING LANDS

ROTATION

UNION CARBIDE MIXER

Fig. 4-9 Mixing pins. Courtesy of Spirex

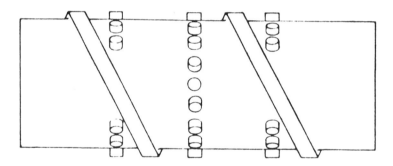

Fig. 4-10 Maddox mixer. Courtesy of Spirex

between. The pins should be hardened and press fit into the screw root to prevent dislodgement.

Pins, unlike other mixing devices, are easy to install as an afterthought. This is usually done after the screw has been used in production and it is found that more mixing ability is needed.

The Maddock mixer (Fig. 4-10), also called the Union Carbide mixer, was patented by Union Carbide and developed for practical use by Bruce Maddock. The patent has been given to the public, so there are no royalty charges.

This screw does an effective job of mixing and screening unmelted material. It has a series of opposed, semicircular grooves along its axis. Alternate grooves are open to the upstream entry; the other grooves are open to the downstream discharge. Ribs, or "flutes," divide the alternating entry and discharge grooves; these also alternate. The flutes are of two types: mixing flutes and wiping and cleaning flutes.

The plastic melt is forced over the mixing flute, which is $\frac{1}{2}$ in. across and undercut about 0.019 in. from the outside diameter of the screw. The cleaning flute is narrower, approximately $\frac{1}{8}$ in., and full diameter. The plastic melt is pumped into the inlet groove, and as the screw rotates, the undercut mixing flute passes under it. The melted material ends up in the outlet or discharge groove. As it goes over the undercut mixing flute, it is subjected to high shear, but for a very short interval. The material is then pumped out of the discharge groove as new material enters over the mixing flute, and it cannot escape over the full-diameter cleaning flute.

Screw Hardening. Screw surfaces, especially the flights and flight lands, are subjected to high frictional forces created during the melting and conveying of the plastic. Some fillers and reinforcements, such as minerals and glass fibers, accelerate the wearing process and can shorten screw life by over 50%. To reduce this wear, the screw can be hardened in several ways.

Flame hardening is one of the oldest techniques used to increase the hardness of the top of the screw flights, which are usually made with AISI 4140 steel. The process uses an open gas-oxygen flame, followed by rapid quenching. The depth of the resulting hardness is about $\frac{1}{8}$ in., and a hardness level of Rockwell C 48 to 55 can be achieved.

Induction Hardening. The result of this process is similar to that of flame hardening. The high heat is created by induction, using magnetic flux reversals, rather than with a flame.

Nitriding. A very hard outside case can be achieved by subjecting the screw or barrel to a high-nitrogen atmosphere at elevated temperatures of about 950 °F. The depth of the case hardening is about 0.020 to 0.024 in., and hardness is in the range of 60 to 70 RC.

Stellite is a product of the Cabot Corporation. When incorporated on the flight lands, it increases the life of the screw by as much as 400%. Stellite material is literally welded onto the top of the screw flight land or into a V-groove cut into the flight land (Fig. 4-11). After the weld bead of Stellite is applied, the flight land is ground to proper size. The resulting thickness of Stellite is about $\frac{1}{16}$ in.

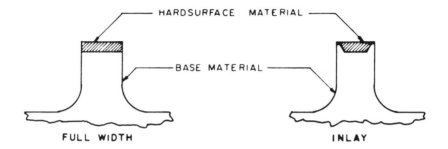

HARDSURFACE MATERIAL

BASE MATERIAL

FULL WIDTH   INLAY

Fig. 4-11 Stellite material welded to screw flight lands. Courtesy of Spirex

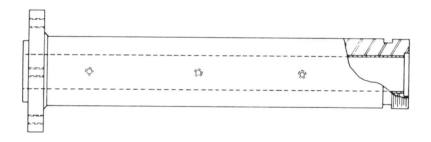

Fig. 4-12 Extrusion barrel. Courtesy of Spirex

The Extrusion Barrel

Extrusion barrels (Fig. 4-12) are often significantly longer than injection molding machine barrels because of the high L/ D ratios used in the production of extruder screws. Instead of having an integral feed port, extruders often have a separate feed area casting, to allow the throat to be cooled more effectively. At the discharge end of the barrel there is a mounting flange to allow the barrel to be fastened to the die adapter.

Today, both extruder and injection barrels are commonly manufactured using a bimetallic material that is highly abrasion resistant. One of the more

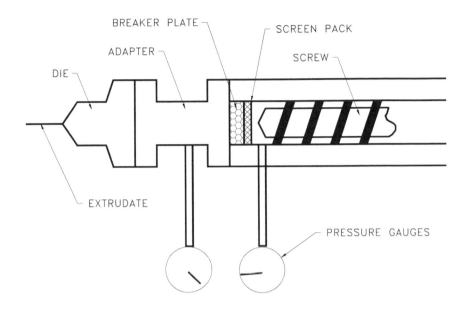

Fig. 4-13 Location of the screen pack and breaker plate in front of the discharge end of the extrusion barrel

popular materials is Xaloy 101, a product of Xaloy, Inc. The metallic components are usually iron and boron. It is possible for the entire barrel to be produced with these alloy metals, but more often only a barrel liner is made from these materials. This lowers the cost and allows for easy replacement when required.

The Screen Pack and Breaker Plate

The screen pack and breaker plate are located directly in front of the discharge end of the barrel (Fig. 4-13). The purpose of the screen pack is to filter out contamination that may enter the melt stream, such as paper and tape, tramp metal (such as box staples), and burned and degraded plastic. If this contamination is not filtered, it will cause defective product and might damage the extrusion die. The screen pack usually consists of several screens that vary in mesh size. The finest mesh screen is placed toward the screw discharge end, and the most coarse mesh size is positioned toward the breaker plate.

As the screen pack becomes clogged, the pressure in front of the screw increases. This provides unwanted back pressure that restricts the passage of the plastic melt and ultimately increases the temperature of the plastic melt. Many extrusion processors place a pressure gauge on either side of the screen pack/breaker plate assembly. Increasing differential pressure is an indication to change the screen pack.

For smaller extruders and short-run production, the screen pack is removed by removing the die and breaker plate assembly, but this method interrupts production. An alternative is to incorporate an automatic screen changer. Automatic screen changers are basically shuttles that hydraulically shift a new screen pack into the place of the old one without interrupting production (Fig. 4-14). Once out, the clogged screen system can be easily replaced, and the shuttle station is prepared for the next screen change.

The breaker plate is a hardened steel plate that nests in the screen pack and provides a controlled amount of back pressure for the extrusion process. Holes in the breaker plate allow the filtered plastic melt to pass through to the die. These holes are streamlined; they are tapered on the side toward the discharge end of the screw. The plastic melt has a turbulent flow as it comes off the end of the screw, but as it passes through the screen pack and the breaker plate, its pattern becomes more laminar. The laminar flow pattern is required for proper flow through the adapter and die, and ultimately for production of a quality extruded product.

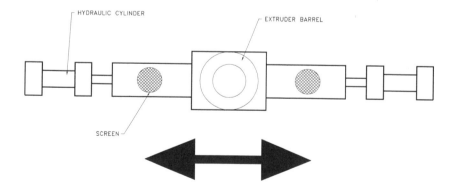

Fig. 4-14 Automatic screen changer

For example, as plastic melt is extruded into a sheet, it is important that the rate of extrusion be consistent across the entire die width. If the plastic exiting the die is flowing in a laminar pattern and at the same rate, the sheet will be uniform in dimension, with minimum curling at the outer edges of the extrudate. If the plastic melt exiting the die is not laminar in nature, or if the material is flowing at different rates across the width of the die, the extrudate will curl and twist at the outside edges, creating a defective product.

Gear Pumps

Gear pumps are used by extrusion processors to improve the output quality and rate of existing equipment. A gear pump is a positive displacement device that exhibits very little slip at high viscosities and provides mechanical pumping efficiencies approaching 95%. Compare this to the pumping efficiency of about 20% for a screw at high pressures.

The principle of operation of a gear pump is simple, but it requires precise mechanical accuracy as well as control synchronization (Fig. 4-15). A gear pump accomplishes its positive displacement by filling each gear tooth with polymer at the entry of the pump (Fig. 4-16), then carrying that polymer in the

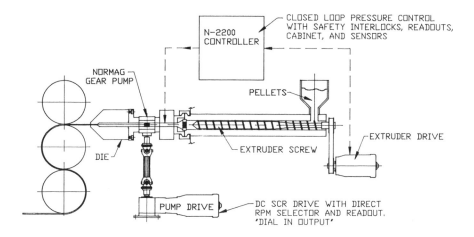

Fig. 4-15 Operation of a polymer gear pump. The extruder melts plastic and feeds it to the gear pump at about 600 to 1000 psi. The controller measures pump suction pressure and automatically adjusts the extruder speed to compensate for varying extruder output. The gear pump delivers plastic to the die at the required pressure to 5000 psi with a volumetric (throughput) accuracy of 1% or better. Courtesy of Nomag

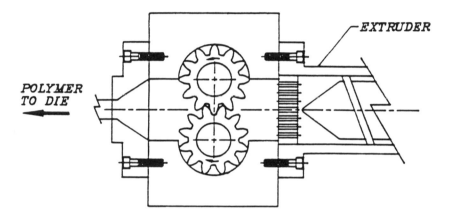

Fig. 4-16 Internal mechanism of a gear pump. Courtesy of Nomag

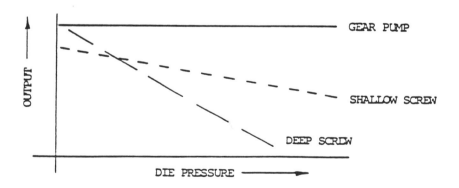

Fig. 4-17 Gear pump output variation as a function of pressure. Courtesy of Nomag

counterrotating cavities of the gear teeth to the pump discharge. As the gears mesh, they provide a rotating seal that prevents the polymer from flowing back to the pump section.

Due to the precise fits and close tolerances of the rotating cavities, very high pressures are achieved at the pump discharge, with extremely good metering of the plastic melt. The pumps are typically powered by direct current drives that determine the system output. The extruder speed is controlled by sensing pump suction pressure and adjusting extruder speed continuously. These precision controls are designed to close the loop and monitor the process continuously.

While extruder screws must slip in order to pump, by definition, they will always have the characteristic of output variation versus pressure illustrated in Fig. 4-17. This graph also shows that pump output is virtually independent of pressure over its operating range for high-viscosity materials. Figure 4-18 compares the typical extrusion or Archimedian screw with the gear pump.

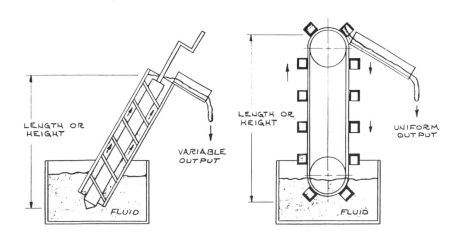

Fig. 4-18 Principle of operation of a gear pump vs. that of an extrusion screw. (a) Extruder screw. Output depends on fluid viscosity, screw speed, closeness of fit, and length. (b) Gear pump having buckets of equal size and constant speed. Output is extremely uniform and does not depend on fluid viscosity, density, fit, or length. Courtesy of Nomag

The gear pump is an excellent tool for improving the efficiency of an extruder:

- It restricts output variation to less than 1% under steady operating conditions, providing that the extruder melts and mixes the material properly.
- It reduces the required pressure at the extruder discharge to as low as 600 psig (psi as read on the gauge), provided that this reduction does not deter the melting capabilities.
- It reduces energy requirements, due to more efficient pumping action.
- It simplifies synchronization with takeoff equipment, due to the linear output of the pump over the speed range.

However, the gear pump is not a panacea:

- It does not improve the plasticating capacity of a poorly designed or marginal screw. The extruder screw must melt and mix the plastic.
- It does not cool the polymer, because the polymer melt residence time is very short. However, by reducing extruder back pressure, a lower melt temperature may be realized.
- It does not improve additive dispersion or mixing. The material must be mixed and melted properly prior to entry into the pump.

Tooling for Extrusion

Tooling for extrusion usually means the dies that form the extruded product.

Die Characteristics. Although dies vary in size, shape, and format, there are some common considerations.

Die Swell. Basically stated, in die swell the extrudate is larger than the die orifice from which it came. Die swell is caused by the rapid decrease in pressure as the plastic exits the die. The polymer molecules are temporarily oriented in a laminar flow within the die, and when the plastic exits the die, the molecules tend to relax to a less oriented state. The die designer and die maker must fashion the die dimensions so that the extruded part will be dimensionally correct.

Flow surfaces, those surfaces within the die that contact plastic melt, must be very smooth. Often the flow surfaces are polished to a mirror-like finish. Additionally, the flow surfaces are often hard chrome plated to resist both corrosion and erosion caused by the plastic melt.

Heat. Dies are usually heated with electric heater cartridges or heater bands. If the plastic melt becomes too cool within the die, it will solidify.

Special Features. Extrusion dies are usually designed with adjustment features that allow the processor to make small changes in extrudate thickness or flow rate as required (Fig. 4-19). Many dies also have a "jack bolt" system to facilitate die disassembly. When a production die is removed from the extruder to be cleaned or repaired, the cooled plastic that remains within the die acts like an adhesive, and the cooled die becomes difficult to disassemble. To prevent this, the jack bolts are tightened after the body bolts have been removed, and because the lower die half is not tapped, the jack bolts push the die halves apart.

Die design can be divided into two main categories, sheet dies and multilayer dies.

Sheet Dies. Two of the most common sheet die designs are the T-type and the coathanger type (Fig. 4-20). Both designs allow the plastic melt to enter the die after the extruder has melted and pumped the plastic into the die. The dies change the flow pattern of the plastic melt to a more laminar pattern and spread the plastic melt pattern to meet the size of the die opening. A deckle, an adjustment on the die, may be used to alter the width of the extruded product.

Multilayer Dies. The extruded products designed today take advantage of the synergistic effect of several combined layers of plastic, formed by using several extruders to melt and pump each plastic into a common die (Fig. 4-21). These complex dies allow extrusion processors to incorporate the individual attributes of different plastics into one product. They also allow processors to sandwich regrind or recycled plastic between exterior layers of virgin plastic. This can lower the overall product cost, and it is an environmentally sound strategy.

Types of Extruded Products

Plastic film comes in several forms and formats that range from the common low-density polyethylene to the precise aluminized polyester film used in modern packaging (Fig. 4-22). The definition of plastic film varies, depending on its quality and application. Most manufacturers agree that film is less than 0.010 in. thick and that if one were to hold an $8\frac{1}{2} \times 11$ in. piece of film at the short end, it would droop, unable to support its own weight. Other definitions of film attempt to differentiate between the relatively low thickness tolerances of high-volume blown film, used for plastic bags, and the tightly

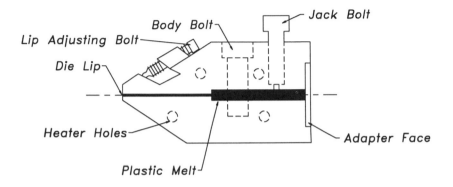

Fig. 4-19 Sheet extrusion die components

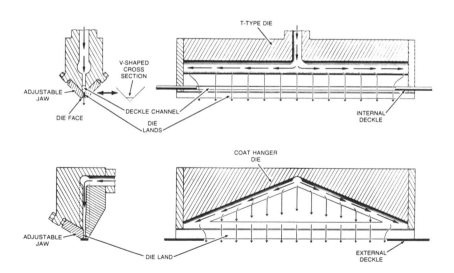

Fig. 4-20 Schematic cross sections of T-type and coathanger-type extrusion dies. The locations of internal and external deckles are indicated. Courtesy of Quantum, USI Division

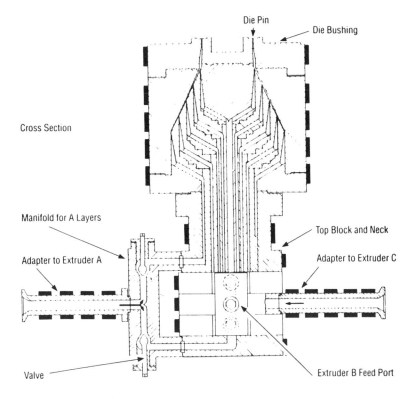

Fig. 4-21 Schematic of a typical five-layer blown film die assembly. Courtesy of Quantum, USI Division

Fig. 4-22 Plastic film products. Courtesy of Quantum, USI Division

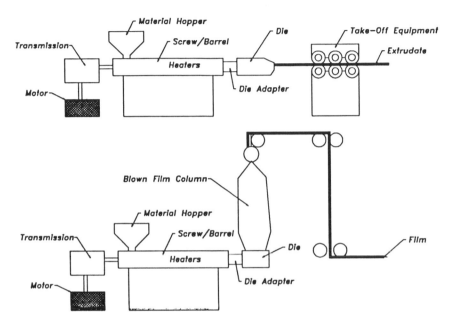

Fig. 4-23 (a) Extrusion of plastic sheet or profile. (b) Extrusion of blown film

controlled thickness tolerances of cast film, used as the carrier for computer printer ink ribbons.

Plastic sheet is similar to plastic film in that it can be extruded in much the same way (Fig. 4-23). To differentiate sheet from film, most manufacturers agree that plastic sheet is over 0.010 in. thick and that an $8\frac{1}{2} \times 11$ in. piece of plastic sheet held at its short end would stand out, supporting its own weight. Typical extruded sheet products include some clear plastic glass replacements and sheet that is subsequently thermoformed into products such as pickup truck bed liners, cups, plates, trays, and refrigerator liners.

Profile is defined as a two-dimensional shape that is extruded in the third dimension. It may range from something as simple as a rod to a very complex geometry (Fig. 4-24). Typical extruded profiles include PVC pipe, PVC house siding, and door and window seals.

Co-extruded products can be film, sheet, or profile. However, instead of using one plastic resin, several plastic resins are extruded simultaneously. This is accomplished by having multiple extruders, each processing one material,

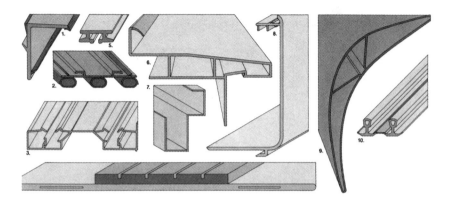

Fig. 4-24 Plastic profiles. 1, 2, and 3, Door and window weatherseals. 4, Roof-deck expansion joint. 5, Windshield wiper spline. 6, Window frame component. 7, Side brick mold for window. 8, Office partition baseboard wire manager. 9, Drift eliminator blade for industrial cooling tower. 10, Hinge for wood panels of folding door. Courtesy of Crane Plastics

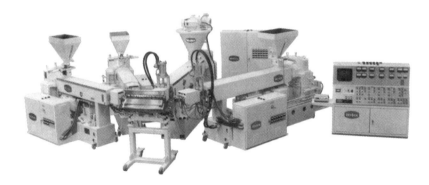

Fig. 4-25 Co-extrusion equipment. Multiple extruders feed plastic melt into a common die. Courtesy of Welex, Inc.

feeding plastic melt into a common die (Fig. 4-25). The co-extruded products can achieve properties that are superior to those of single-material extruded products, or that are unavailable from single-material extruded products. This synergistic effect supports the design and manufacturing of single plastic parts

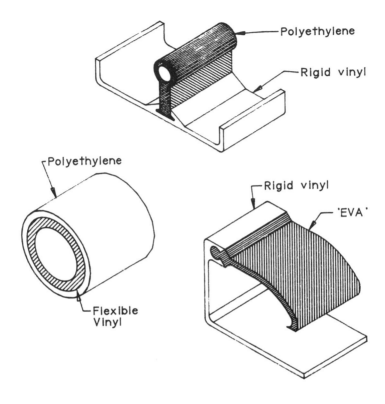

Fig. 4-26 Nonbondable materials joined by mechanical keying or fit

that have the properties of products that would otherwise have to be assembled (Fig. 4-26 and 4-27).

Filaments. Filaments are thin strands or threads produced by using an extruder to pump plastic melt through a filament or strand die that produces several strands at one time. The filaments must be immediately cooled in a water tank to prevent them from sticking to each other. Examples of extruded filaments are nylon fishing line and nylon and polyethylene lawn trimmer line.

Coatings. The extrusion process can be employed to apply a layer or layers of plastic melt, usually low-density polyethylene, onto a substrate, improving

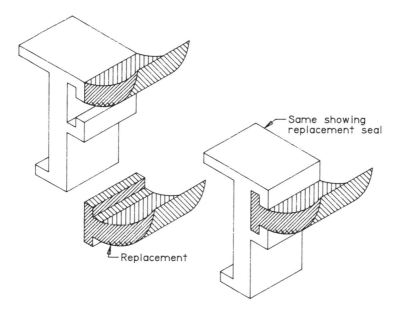

Fig. 4-27 Bondable materials. If the flexible sealing portion wears out, a replacement flexible insert can be placed into a slot in the rigid profile.

the properties of the substrate. A plastic coating on materials such as paper, paperboard, cloth, metals, and glass provides improved weather resistance, water resistance, and overall durability. The coating thickness usually varies from 0.00025 to 0.004 in.

To receive a plastic coating, cardboard or any other substrate must be fed under an extrusion die and must subsequently be gauged or sized using rollers and trimmers (Fig. 4-28 and 4-29). The extruded plastic is deposited between a pressure roll and a chill roll, through which the substrate to be coated is passing. (This portion of the coater is often referred to as the laminator, but the term coater is more descriptive.) Aside from the extrusion system, most of the rest of the equipment is associated with unwinding the substrate and winding the coated substrate.

The pressure roll is a large metal idler roll with a thick, hard covering of neoprene or silicone rubber. To prevent the plastic melt from sticking to the pressure roll, a water-cooled chill roll is used. It has three main functions:

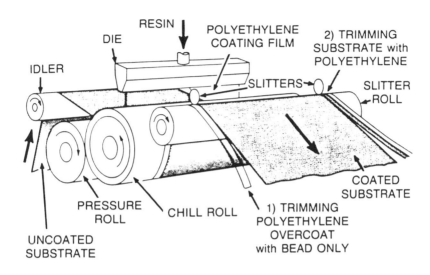

Fig. 4-28 Extrusion coating. Either polyethylene "overcoat" with "bead" (1) or polyethylene-substrate scrap (2) is trimmed. Both slitters do either one or the other. Courtesy of Quantum, USI Division

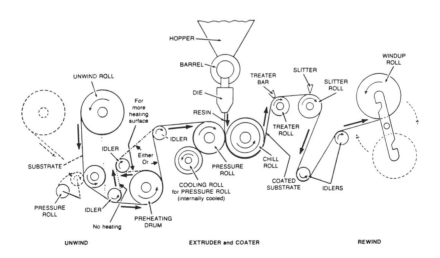

Fig. 4-29 Schematic cross section of an extrusion coating line, with unwind and rewind equipment. Courtesy of Quantum, USI Division

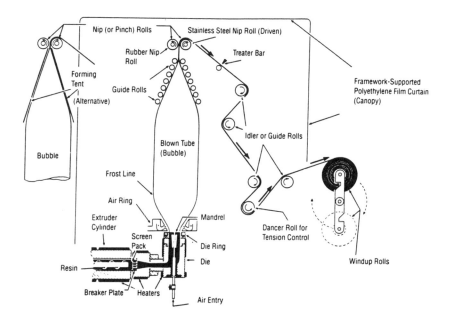

Fig. 4-30 Blown film extrusion. The blown film bubble rising up from the die is pulled into a complex takeoff system. Courtesy of Quantum, USI Division

- To cool the coated substrate to allow it to be stripped
- To help control the coating thickness
- To determine the surface quality of the coating

Special Processes

Blown film extrusion of plastic (predominantly of low-density polyethylene) involves feeding the plastic melt into a ring-shaped die, through either the bottom or the side. The melt is forced around a mandrel inside the die, shaped into a sleeve, and extruded through the die opening in the form of a comparatively thick-walled tube. While still in the melt state, the tube is expanded to a bubble or hollow cylinder of the desired diameter and corresponding lower film thickness (Fig. 4-30). This expansion is accomplished by the pressure of

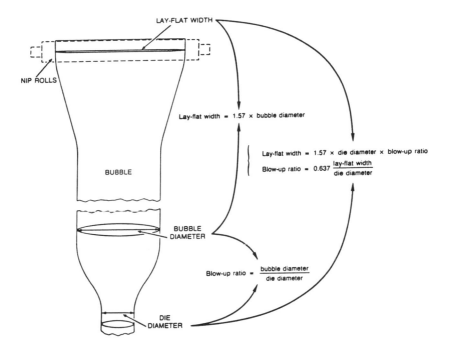

Fig. 4-31 Schematic drawing of a blow-up ratio of 2.5 (bubble diameter divided by die diameter) and the lay-flat width after the nip rolls have flattened the bubble to a double layer of film. Courtesy of Quantum, USI Division

internal air admitted through the center of the mandrel. Once the bubble has been formed for the desired lay-flat width, no additional air is required to keep the bubble and its diameter (and thereby the blow-up) stable.

After a few yards of free suspension, the bubble is flattened between two nip rolls, and it is ultimately festooned through a series of rollers to the wind-up rolls at the end of the processing line. The frost line in Fig. 4-30 refers to the ring-shaped zone where the bubble begins to change from a clear melt, which represents the heated amorphous molecular structure of the plastic, to the frosty appearance of the cooler melt, which represents the semicrystalline nature of the plastic. The height of this frost line represents the result of the control of the molecular orientation and ultimately the physical properties of the film. Figure 4-31 is a closer view of the bubble-die relationship.

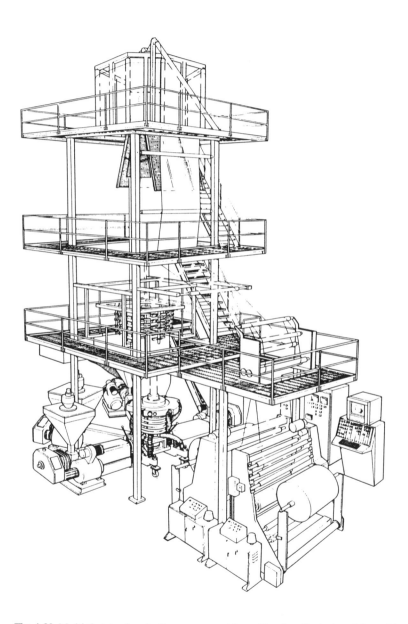

Fig. 4-32 Multiple extruders feeding a common blown film ring die, designed for making multilayer film. Courtesy of Quantum, USI Division

More complex blown film lines, now being developed, take advantage of the synergistic properties of multiple layers. These layers are produced by having several extruders feed into a common blown film ring die (Fig. 4-32). The resulting film product is often a tear-resistant, high-strength plastic that can be used for heavy-duty trash bags or packaging materials.

Cast film extrusion allows the plastics processor to manufacture a film product with better tolerance control than is possible in the blown film process. The extruder used in the cast film process is a standard flat-film unit with either a T-type or coathanger-type die, but the chilling and takeoff equipment is more sophisticated than in other extrusion processes. In addition to superior dimensional control, cast film provides better clarity and gloss than conventional blown film does.

The plastic melt, usually low-density polyethylene, is extruded through the die slot and cooled by the surface of two or more water-cooled chill or casting rolls (Fig. 4-33). The hot polyethylene web drops onto the first chill roll it contacts tangentially, so alignment of this roll in relation to the falling film is critical. To pin the molten plastic to the chill roll, a gentle stream of air from an air knife is often used. The chill rolls are highly polished, chrome-plated cylinders with precise dimensions. The plastic is S-wrapped around the chill rolls and across a series of takeoff rolls to the wind-up station. The thickness or

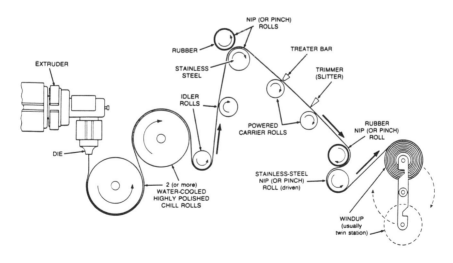

Fig. 4-33 Schematic drawing of chill-roll film extrusion equipment. Courtesy of Quantum, USI Division

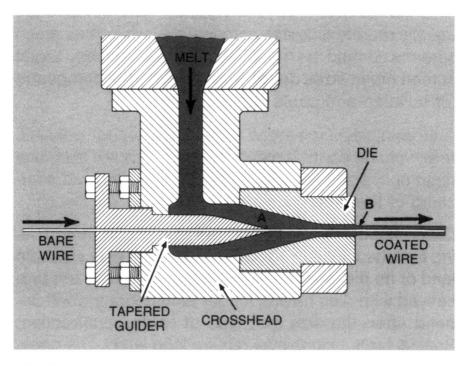

Fig. 4-34 Wire coating. A crosshead holds the wire coating die and the tapered guider. Courtesy of Quantum, USI Division

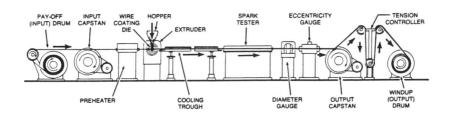

Fig. 4-35 General layout of a wire coating extrusion plant. Courtesy of Quantum, USI Division

gauge of cast film is controlled by balancing the die opening; the chill roll size, spacing, and temperature; and the take-off speed.

Wire coating is the process of extruding plastic wire insulation material(s) over metal wire in a continuous operation. The extruders used in wire coating are not unlike those used in other forms of extrusion. The key difference is that the wire coating die is held in a crosshead that also holds the tapered guides mounted axially with the die (Fig. 4-34). The extruder screw forces the plastic melt down and over the guides through which the wire is drawn, and out through the die in which the coating forms around the wire. The die from which the sheathed wire emerges is usually provided with a land or, for thin coatings, with a knife edge. The inlet to the die has streamlined contours to increase coating speed.

Coating thicknesses vary from 5 mils to more than $\frac{1}{2}$ in. The coating operation of the die is often measured by its draw-down (the ratio of the cross-sectional area through which the plastic melt is extruded to the cross-sectional area of the finished coating). Today's wire coating processes use several plastic and rubber materials, including polyethylene, cross-linked polyethylene, PVC, rubbers and elastomer, nylon, and fluorocarbons.

Figure 4-35 illustrates the general layout of a wire coating extrusion line. Wire, usually copper stranded or solid, is fed to the wire coating die where, as described earlier, it is coated with the plastic melt. After cooling, the coating wire is tested for its insulation quality using an in-line spark tester. The spark tester is a nondestructive test that employs high voltage and low current to inspect for pin holes and other insulation quality defects. Later in the process the wire is inspected for its mechanical dimensions, such as outside diameter and eccentricity. Prior to being wound on a spool or coiled in a drum, the wire is coded with an in-line ink-coder or laser system.

Future Applications

The extrusion process is the workhorse of the plastics industry and will continue to be adapted to other processes and materials. The biggest growth will be in short-run, custom applications. Historically thought of as a continuous, high-volume process, extrusion is now being adapted to allow quicker set-ups and materials changes, facilitating the production of more varied short-run plastic products.

5

Blow Molding

Blow molding is the most popular and productive process used to form hollow plastic parts (Fig. 5-1). Initially, it was almost exclusively used to produce plastic containers such as bleach bottles and milk jugs. The plastic containers were an immediate success because they replaced heavy glass containers that could easily break. As more plastic materials were developed for the blow molding process, glass container production fell and plastic container production skyrocketed. In the early 1970s, plastic materials were developed that would minimize gas transfusion. These "barrier plastics" were used for carbonated soft drink containers, again displacing glass.

As plastic bottles and other containers became more sophisticated in their design and processing requirements, processing technology and equipment improved. Eventually, blow molded parts were developed to replace metal parts in the area of fuel and chemical tanks. Today, the blow molding industry is growing at a rapid rate as larger, more complicated plastic parts are being produced (Fig. 5-2). This evolution has resulted in the development of two distinct processes, extrusion blow molding and injection blow molding.

Fig. 5-1 Blow molding equipment. Courtesy of Quantum, USI Division

Extrusion Blow Molding

Extrusion blow molding systems use an extruder to form a tubular extrudate called a parison, which is still a plastic melt (Fig. 5-3). Before the parison can cool and solidify, the blow mold traps it, sealing both ends. Immediately after the parison is captured by the mold and sealed, a blow pin is inserted into the top of the mold, and shop air at 100 psi is forced into the tubular parison. (A blow pin allows passage of blowing air from the air manifold into the containers. In many systems it is also used to exhaust air out of the container once the blow cycle is complete.) The air pressure causes the parison to balloon and take the shape of the inside of the mold. After the plastic melt cools and solidifies in the closed mold, the mold opens and the blow molded part is ejected.

Fig. 5-2 Plastic products produced with blow molding. Courtesy of Quantum, USI Division

Extrusion blow molding uses either a continuous or intermittent basis for the formation of the parison. Both systems have advantages, depending on the type of container to be molded, the speed at which the process is to be run, and the size of the part. For example, extremely thin-walled beverage containers need an accumulator-type continuous extrusion system. Consideration should be given to all these factors before deciding on the purchase of a new blow molder for a particular type of product line.

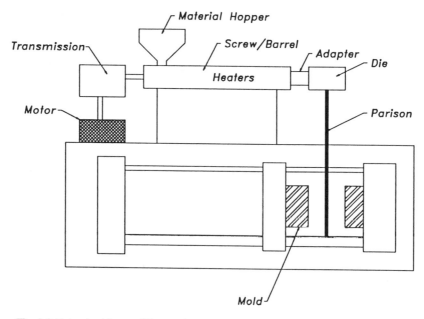

Fig. 5-3 Extrusion blow molding equipment

To clarify a somewhat confusing point of terminology in regard to continuous versus intermittent extrusion, it should be said that theoretically, accumulator-type machines are continuous, in that extrusion into an accumulator reservoir is constantly taking place. Parison dropping, however, is intermittent.

Intermittent Parison Formation

A variety of intermittent parison drop systems are available. The most common is the single-plane-mold-motion reciprocating screw type, widely used for containers in the size range of 32 oz to $2\frac{1}{2}$ gal. Half-gallon and gallon containers are almost exclusively produced on this type of equipment.

The reciprocation process is fairly simple, as the press section onto which the mold halves are mounted (referred to as the platens) moves in only one plane. Parisons are dropped between the open mold halves, the mold closes on them, and then air is blown into the parisons to form the containers. During the blowing cycle the screw to the extruder turns, forcing plastic melt

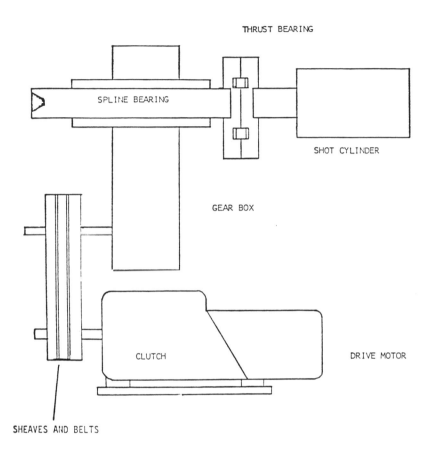

THRUST BEARING

SPLINE BEARING

SHOT CYLINDER

GEAR BOX

CLUTCH

DRIVE MOTOR

SHEAVES AND BELTS

Fig. 5-4 Reciprocating screw system

to be advanced in front of it. This action forces the screw backward. Once the blowing cycle is complete, the mold halves open and an electrical signal is sent to the hydraulic cylinder to push the screw forward, delivering the parison to the next drop. The screw is constantly rotating during automatic operation of the machine. A typical reciprocating screw system is illustrated in Fig. 5-4.

Reciprocating Screw System Components. The key components of the reciprocation process are the spline bearing system, the thrust bearing, and the shot cylinder.

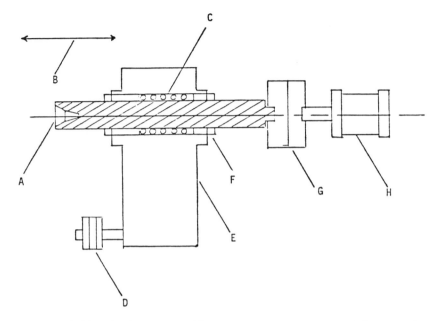

Fig. 5-5 Detail of the spline bearing assembly in a reciprocating screw system. A, Screw coupling. B, Direction of reciprocation. C, Spline bearing. D, Drive sheave. E, Gear box. F, Oil seal. G, Thrust bearing. H, Shot cylinder.

The spline bearing system (Fig. 5-5) is typically a cylinder housing a number of large ball bearings. The entire system is contained within the hollow center section of the gear reducer. In essence, the spline bearing is necessary to minimize the friction and drag created in the reducer by the screw reciprocation, at the same time that it is rotating under full torque. If properly lubricated, the entire gear reducer/spline housing assembly will be trouble-free.

Thrust Bearing. The pressure developed by the mass of plastic melt being pumped through a small die opening is also transmitted to the screw, which pushes the screw against the thrust bearing. The thrust bearing consists of a series of rollers mounted in a doughnut-shaped casing. The rollers are positioned in the casing somewhat like the numbers on a clock face. The entire casing is then sandwiched between polished steel races, which are enclosed in a two-piece housing. The back half of the housing is coupled to the front of the shot cylinder shaft and does not rotate during reciprocation. Usually the front half of the housing is coupled to the back end of the split bearing inner shaft.

This inner member is, in turn, coupled to the back end of the screw. Thus, as the screw turns, so does the inner member of the spline bearing and the front half of the thrust bearing.

Shot Cylinder. Usually sized according to the extruder size, the shot cylinder is a nonrotating hydraulic cylinder that pushes the screw forward for shot delivery. For example, a $3\frac{1}{2}$ in. extruder would generally be equipped with an 8 in. bore cylinder, the stroke of which would be 6 in. or more. Some machines vary from this standard in that the shot cylinder shaft rotates with the screw and no thrust bearing is used. Basically, however, some type of pushing cylinder is needed in all reciprocating screw machines. It must be large enough to exert sufficient pressure to drive the screw and melt forward and deliver the proper amount of shot for the size and weight of the parison being blown.

The length of stroke of the shot cylinder and the pressure at which parisons are dropped are adjustable within certain limitations. Theoretically, the length of shot stroke should be adjustable from zero to the full capacity of the cylinder length. Cylinders are generally oversized, however, and the extruder's ability to plasticate only a limited amount of resin per hour usually restricts the full length of the shot stroke from ever being used. Plasticating capacity is determined by the size of the extruder, the screw design, the resin type, and the screw speed at which the extruder can run at maximum. This speed is controlled by the motor size.

Parison Diameter. Intermittent parison production presents the greatest number of problems with parison production and control, making operator attention much more intensive than with continuous parison drop operations. Parison flare, a processing characteristic emphasized by intermittent parison drop, is critical to both tooling design and day-to-day parison production. Simply defined, parison flare is the increase in diameter of the parison as it leaves the die opening.

Particularly in the production of handled containers, the parison must be large enough to effectively fill the mold cavity while maintaining a blow-up ratio of not greater than 4 to 1. (Blow-up ratio is the relationship between the size of the parison and the size of the finished container. For example, if the finished diameter of a round gallon container with a handle is to be 6 in., the parison diameter must be at least $1\frac{1}{2}$ in., with 2 to 3 in. being more suitable.) The diameter of the parison is determined by the die and mandrel design used. Figure 5-6 shows the three basic die/ mandrel configurations:

- A *straight die/straight mandrel* (Fig. 5-6a) results in little flare, but it also results in a fixed die gap, so that the parison wall thickness cannot be changed.

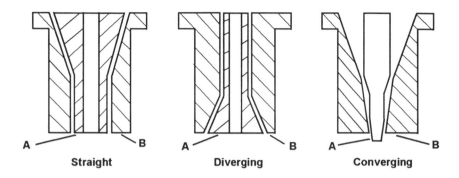

Fig. 5-6 The three basic die/ mandrel configurations for intermittent parison production. A, mandrel; B, die

- *Diverging tooling* (Fig. 5-6b) is very widely used in the production of thin-walled, handled containers. This design illustrates flare at its best. As the melt flows downward toward the die opening, the angles in the die and mandrel change its direction. Once the melt exits from the die opening, it tends to follow the path set by the angle of the mandrel until it collects enough of its own weight that gravity pulls the whole mass inward.

- *Converging tooling* (Fig. 5-6c) is arranged so that the mandrel extends well below the face of the die. It thus guides the parison on a straighter downward path and reduces the tendency of the parison to collapse on itself.

In observing parisons being dropped on an intermittent parison machine, visual discovery of the flare phenomenon requires close attention to the die tip area. The parison drop is generally quite fast, and because the flaring action is most noticeable in the first inch or so of extrudate, it may be visible for only a fraction of a second.

In order to produce the proper amount of flare for the parison size desired, die and mandrel diameters can be varied significantly. It is also possible to vary the angle cut into the mandrel, especially at the tip, and correspondingly into the die. A typical example is milk gallon tooling, which is quite standard industry-wide. The desired parison has a diameter of approximately $3\frac{1}{2}$ in.; the diameter of the tooling to produce this parison is $2\frac{3}{4}$ in.

Intermittent Drop Machine Components. Intermittent parison drop machines generally evolve around a basic extruder that could just as well have been adapted to tube or profile extrusion, film extrusion, or even injection molding. What vary are the head section and the press section.

Head Section. The custom-fabricated head section of any blow molder is really what machine designers engineer and develop, and it is what many blow molding machine manufacturers actually patent. Intermittent machines can have anywhere from one to twelve individual heads tied together by a common manifold (Fig. 5-7). The head section channels the flow of the resin melt, drops this melt through the die, and thus forms parisons. Integrally connected to the head section is what is commonly called the air manifold system, which allows blown air to inflate the parison in the molds, thus forming containers. The entire head section on both intermittent and continuous parison drop machines is fairly straightforward. It is the inner members of these heads that differ.

Press Section. Also located forward of the extruder, and perhaps the most variable of all components with respect to individual machine manufacturers, is what is generally referred to as the press section. As with injection molding machines, the press is what effects the opening and closing of the molds and the clamping forces necessary during the blowing cycle to keep the mold halves sealed tightly together. This section is often called the clamp section for this reason. In general, mold halves are bolted to parallel thick steel plates called platens. Usually these platens are tied together and held parallel by four, six, or even eight steel rods called tie rods or tie bars. The rods are often threaded on at least one end and held in place by means of large nuts.

Continuous Parison Formation

The second type of extrusion blow molding system, called the continuous parison drop, is possibly the more popular in terms of the total number of machines operating worldwide. Particularly effective in the production of medium-size nonhandled containers, this type of machinery emerged with the growth of plastic packaging in the household products market. In recent years, penetration of this equipment into the handleware field has been extensive, so that today nearly any container, from 4 to 128 oz, can be produced on a continuous parison machine.

The primary question that should be asked in comparing continuous to intermittent parison drop is, "What is the advantage of one over the other?"

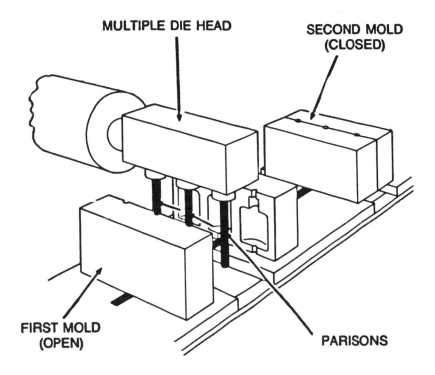

Fig. 5-7 Sliding-mold extrusion blowing with a multiple-die head. Courtesy of Quantum, USI Division

The answer can be simple or complex, depending on one's particular preferences. Continuous parison drop offers fewer parison control problems, because parisons much smaller in diameter can be dropped for producing the same type of container. Varying the parison thickness at desired points through automatic parison programming can be more easily achieved, because the parisons are extruded at a much slower rate than in intermittent extrusion. Coupled with the fact that extrusion is taking place during the blowing cycle, this means that no time is lost waiting for parisons to drop.

On multiple-station turntables, described below, a large number of cavities can be used to maximize productivity. On other types of machines, container production per hour has generally been lower, given the same number of cavities, the same container weight, and equivalent mold cooling efficiency. The choice of process depends on the type, size, specifications, and number of containers to be run, as well as each individual's manufacturing objective.

Continuous Drop Machine Design. As previously stated, the major difference between intermittent and continuous parison drop systems lies in parison production. In the continuous system, parisons are constantly being extruded as long as the extruder is running. Where the parisons fall during the blowing cycle, relative to the position of the mold mounting, varies with individual machine types. Almost all continuous parison drop machines with multiple head arrangements employ some type of moving mold system. (This statement may be debated in some circles, in that a few fixed mold systems are in operation. For the most part, however, these are older machines that do not achieve the efficiency of the modern equipment on the market.) The moving mold systems have a number of basic designs.

Multiple-Station Turntable. The first design employs a wheel that rotates in a horizontal or vertical plane. The horizontal system is similar to a lazy Susan on an enormous scale (Fig. 5-8). A single parison is constantly extruded vertically from a fixed head, and a series of molds mounted on the indexing table rotate and stop individually under the head. A length of the exposed parison is then secured and cut off by the closing mold, and the wheel rotates again while the blowing portion of the cycle begins. This process continues through a series of stations until the container is sufficiently cooled to be ejected at the station just prior to the one positioned under the head.

The vertical wheel, similar in design to a Ferris wheel, operates in basically the same manner (Fig. 5-9). Historically, a large number of molds have been mounted on this type of wheel, allowing for greater bottle production per hour. Many of the wheels currently in operation are proprietary, in that they are custom-designed and built by the user and are not readily available on the market.

Shuttle Press. A second type of continuous parison drop machine involves the extrusion of multiple parisons that are captured by horizontally shifting multiple molds (Fig. 5-10). In a typical system, four parisons are being extruded in the spaces adjacent to four molds. The platens and the molds in this case move in two planes. First, the molds open and close front-to-back, as is common to all systems. Second, the entire platen assembly is able to shift left-to-right and right-to-left in a line 90° from the mold open-and-close travel, on a separate set of tie bars. In any given cycle, with the platen assembly shifted

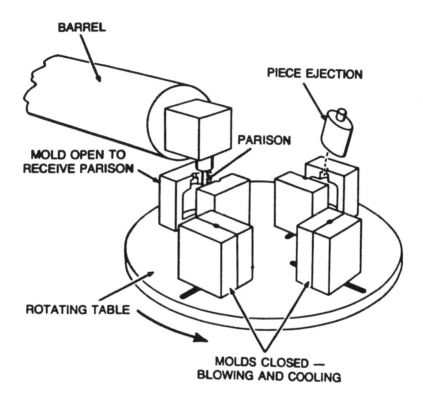

BARREL

PIECE EJECTION

PARISON

MOLD OPEN TO
RECEIVE PARISON

ROTATING TABLE

MOLDS CLOSED —
BLOWING AND COOLING

Fig. 5-8 Multiple-station turntable. Courtesy of Quantum, USI Division

to the right and four containers being blown, four parisons are simultaneously being dropped between the mold blocks.

Once the blowing cycle is complete, the molds open, the containers are ejected, and the entire platen assembly shifts to the left. The molds then close on the hanging parisons, and the platen assembly shifts to the right again. During the second shifting procedure, four red-hot cutoff blades cut the captured length of the parison so that the extrusion of the next set of parisons can continue. Once the platens are completely shifted over, blow pins drop down into the neck blocks, and air begins to inflate the parisons to form the containers.

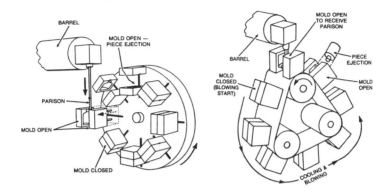

Fig. 5-9 Multiple molds on (a) a vertical rotary wheel and (b) an endless belt. Courtesy of Quantum, USI Division

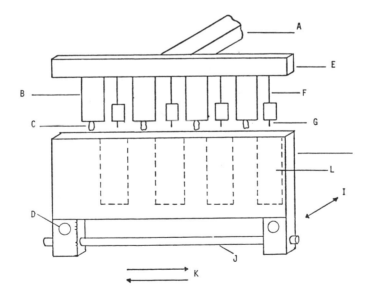

Fig. 5-10 Continuous parison formation shuttle press. A, Extruder. B, Extrusion head. C, Parison. D, Tie rod. E, Head manifold. F, Blowing head. G, Blow pin. H, Platen. I, Mold open-and-close direction. J, Tie rod. K, Shuttle direction. L, Blow mold

In the shuttle press system, necks can be formed by either the conventional flashed neck system or by the captured neck system. Flashed neck systems are generally used on handled containers where the parison has to be large enough to reach the handled area. This system is often used on oddly shaped containers having a large blow-up ratio. The captured neck system is commonly used on nonhandled round or oval containers. Here the size of the parison is the same as the neck thread outside diameter, so no excess flash travels down the parting line on the threads and shoulder of the container. The system allows for a simple knockoff wheel timing system that needs to remove only a small piece of flash at the top of the neck.

A number of variations on the shuttle press system are currently operating in the field. The European influence is seen in machines that allow platen shifting on about a 25° annular plane. Considering the front of the machine as the face of a clock, two sets of platens are visible. While the right-hand platen is at the twelve o'clock position capturing parisons, the left-hand platen is at the eight o'clock position blowing containers. The platens move diagonally across the clock on hydraulically operated arms (Fig. 5-11). An alternative system employs a rising platen assembly that moves up and down on an inclined plane (Fig. 5-12).

Valved Manifold. A third and totally different type of continuous parison drop line operates with a shifting valve at the center of the head manifold (Fig. 5-13). In this system, extrusion is continuous, and flow is diverted alternatively to the right and left heads by means of a valve located in the center of the manifold. While the left head is blowing a container, the valve is open to the right, allowing extrusion through the right-hand die or dies. Once a sufficient length of parison has been extruded, the mold or molds on that side close, the valve opens to the left, and the process continues. This system is classified as continuous parison drop because a parison is continuously being dropped on one side or the other.

Parison programming is perhaps the most significant advantage of continuous parison drop systems. Depending on container design, the need to maintain a particular wall thickness in the areas farthest away from the center of the mold, while maintaining fast cycle times, can become a critical problem. Parison programming can overcome this difficulty by adding material in strategic locations along the length of the parison (Fig. 5-14). This process is accomplished by automatically varying the die gap during the parison drop time.

The heart of this system is an electronic computerized programmer on which a pattern can be preset in normal sequences of several points. The

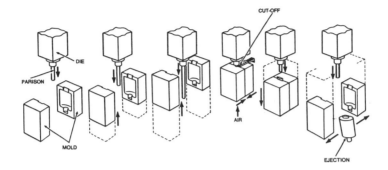

Fig. 5-11 Rising platen assembly for a shuttle press. Courtesy of Quantum, USI Division

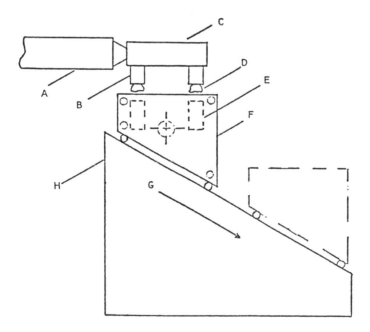

Fig. 5-12 Rising platen assembly for a shuttle press in which the mold travels on an inclined plane. A, Extruder. B, Extrusion head. C, Head manifold. D, Parison. E, Molds. F, Platen in parison collection position. G, Travel direction for blow cycle. H, Frame

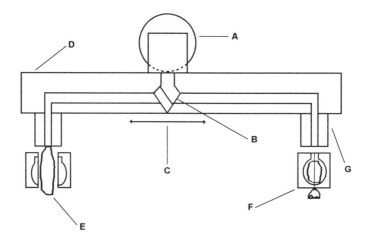

Fig. 5-13 Valved manifold system for a shuttle press. A, Extruder. B, Shifting valve. C, Direction of flow. D, Head manifold. E, Parison. F, Mold. G, Extrusion head

number of points is based on the capacity of the microprocessor attached to the blow molder. By setting the various points on the microprocessor-based parison programmer, electronic impulses are sent to a hydraulic system attached to the mandrels. The mandrels, in turn, raise or drop at the given points during the parison drop, so that material can be added or removed from the parison wall thickness at any point desired. The larger the microprocessor capacity, the more points available. This allows for greater process resolution and control.

Programmed parison profiles usually have to be gradual to be effective. You could not, for instance, change the parison wall thickness by 0.025 in. between points 4 and 5. Such a change in thickness would probably have to begin at point 2 and increase gradually through points 3 and 4 until the 0.025

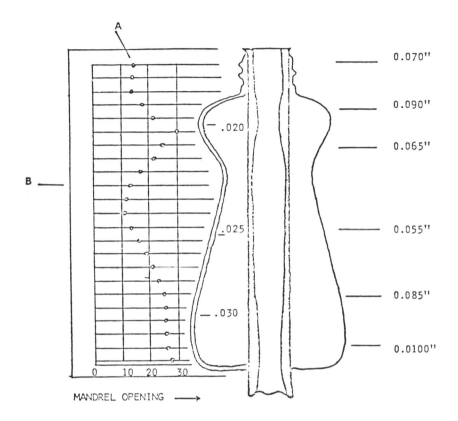

Fig. 5-14 Parison programming. A, Movable program pins. B, Programmer grid

in. increase could be achieved at point 5. This is an important factor to consider when specifying a programmer on a blow molder, especially if the machine is to run a specific container. If the container is intricate in design, with shoulders, indentations, and undercuts, a programmer with at least 32 points may be required.

Not all containers need to be processed with a programmed mandrel, even those of seemingly difficult design. In many cases, proper tooling design can solve a number of parison production problems. Some containers do, however, need to be programmed to be blown in acceptable quality. It is generally agreed that any container can be made better through programming, even if container weight reduction is the only measurable result. Nevertheless, parison programming is not always needed in order to produce a salable container.

Two other types of programming deserve mention here, although they are not as prominent in the field as the electronic system. One system involves mechanically activated cams that are tied in to the mandrel movement and can vary the die gap once or twice for each parison drop. The second involves varying the shot pressure, and thus the swell, through timers tied into the hydraulic system.

Although parison programming has been used in intermittent parison drop systems, its effectiveness is generally less significant because of the extremely fast parison drops desired by most blow molders using intermittent systems. In order for programming to be truly helpful in these systems, parison drop times would need to be slowed significantly, which would increase cycle times and lower productivity. The reason for this is that mandrel movement to create a varying die gap would have to take place in fractions of seconds, as opposed to the 5 to 6 seconds available in a continuous parison drop operation.

The choice of whether to use programming on intermittent parison drop systems depends on individual preference and on productivity versus product results. For example, there are gallon containers being programmed at sacrificed productivity because of the end user's quality specifications. Also consider that sudden contour changes are not easily accomplished, but that for smooth contour changes, programming is the optimum processing tool.

Container Types

A wide range of containers are molded on the various extrusion blow molding systems. Some of the more common designs are depicted in Fig. 5-15. Generally speaking, containers in the size range from 16 to 48 oz, and occasionally 8 oz, are molded on continuous parison drop systems. Lightweight containers from 28 to 64 oz, and occasionally those with handles, are generally run on intermittent parison drop systems. There are several exceptions to these statements.

The common determining factor in selecting a system for a container is economics. Many blow molders feel that cycle time is of the utmost importance, while others feel that tighter control is the key factor. There are a number of excellent blow molding systems on the market that can adequately produce most containers, and selecting the right one is a matter of personal preference and affordability. Container size does present some minor limitations. For

Fig. 5-15 Typical containers produced with extrusion blow molding

example, blow molding 1 oz pharmaceutical containers on a four-head intermittent parison drop machine would be economically prohibitive.

Neck Finishing

A number of methods are available for forming and finishing necks on
blow molded containers. All of the varieties apply to extrusion blow molding,
because injection blow molded necks are formed in the parison mold. The

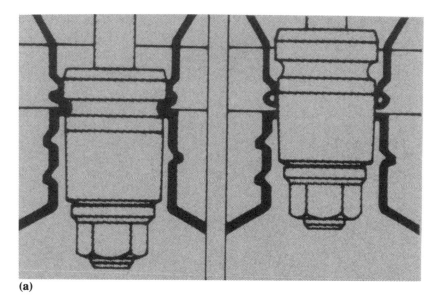

(a)

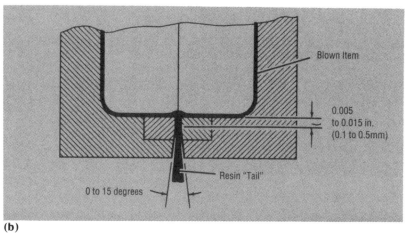

Blown Item

0.005
to 0.015 in.
(0.1 to 0.5mm)

Resin "Tail"

0 to 15 degrees

(b)

Fig. 5-16 The three most common methods for neck finishing. (a) Compacting method. (b) Dome method. (c) Prefinished shearing method. Courtesy of Quantum, USI Division

(continued)

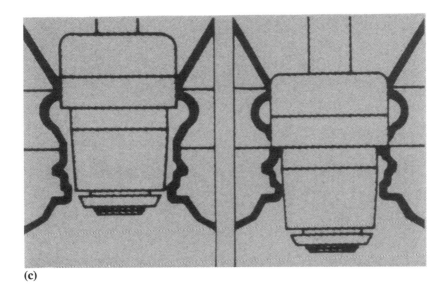

(c)

Fig. 5-16 The three most common methods for neck finishing. (a) Compacting method.
(b) Dome method. (c) Prefinished shearing method. Courtesy of Quantum, USI Division

three most common methods are compacting, dome, and prefinished shearing
(Fig. 5-16). A prefinished compacting method is also being used.

The compacting method (Fig. 5-16a) pneumatically drives the blow pin
down into the neck ring cavity once the mold closes on the parison. When the
blowing cycle is complete, the blow pin is retracted and the container is ejected
to be trimmed by a high-speed rotating blade. The neck is considered prefinished.

The dome method (Fig. 5-16b) was originally designed to allow blow
molding of a container whose neck is not on center. It also allows clamping of
the mold around a stationary blow pin. After the neck ring grasps the parison
away from the head center, the resulting flash resembles a dome and must be
trimmed off. The neck must then be reamed and faced to achieve proper
dimensions.

The prefinished shearing method (Fig. 5-16c) allows the neck ring to close
around a blow pin that is in a down position. When the blow cycle is complete,
the blow pin pulls up and rubs against a close-fitting shear ring. The result is
a completely finished neck that requires no secondary operation, other than

automatically knocking off the surrounding flash in a guillotine trimmer. This system is most commonly used in trimming milk containers.

The prefinished compacting method has the molds close first; then the blow pin is driven down, forcing material into the neck ring while shearing the lip of the neck. A variation of this system allows for blowing the neck finish but still shearing and prefinishing the lip with the downstroke of the blow pin against steel shear rings.

Most blow pins are precision tools in that neck rings clamp around them, creating a seal during the blow cycle. As in the prefinished neck system, the blow pin also determines the quality of the finished neck. Worn blow pins often result in trimming problems, or at least in unacceptable excess flash in the container lip.

Injection Blow Molding

Like extrusion blow molding, injection blow molding is the result of the marriage of two plastics processing systems. The basis for injection blow molding is injection molding, which is the most common of all the systems used to mold solid plastic articles. The molded preform from which the container is blown is formed by injecting molten resin into the preform mold. The resin is plasticated by the reciprocating screw of the injection molder (see Chapter 6).

In both extrusion blow molding and injection blow molding, the screw comes forward to deliver the charge of melt. In extrusion systems, this melt flows out in a parison. In injection blow molding, the melt is channeled through a small opening called a sprue bushing. The melt is then forced under high pressure through a gate and into a mold cavity that is machined to have the shape of the desired parison. However, in the case of injection blow molding, this parison is called a preform (Fig. 5-17).

A number of things are accomplished during this first step in molding the preform. First, if the container is to have threads (which is usually the case), the thread section of the neck has been injection molded to exact tolerance. Second, the even wall thickness of the finished container has been preset by the shape of the preform, providing that the preform mold has been properly designed. Referring again to Fig. 5-17, note that in blowing up a preform of uniform thickness to form the container shown, the wall thickness farthest from the center of the mold would have become extremely thin once the blown air had stretched the material enough to meet the mold walls. By molding the preform with thicker walls, corresponding to the farthest points from the

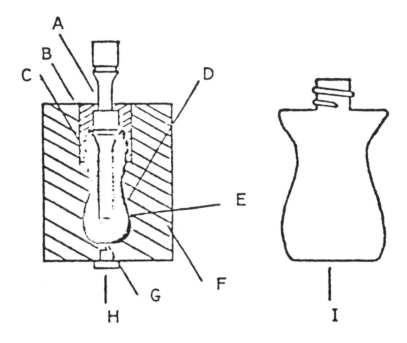

Fig. 5-17 Preform mold. A, Core rod. B, Neck block. C, Molded threads. D, Preform cavity. E, Core rod tip and air entry point. F, Mold. G, Gate. H, Nozzle or sprue bushing. I, Finished container

center on the container, variations in container wall thickness can be virtually eliminated. The end result is that production of fairly intricate containers, which possibly would not otherwise be blow molded, becomes a relatively simple task.

These two major advantages, coupled with the fact that injection blow molded containers generate no flash to reprocess, are strong selling points for this type of blow molding. Some of the newer safety-type necks, because of extremely close tolerances, could not be blow molded in any other way.

Process Machinery

A typical injection blow molding system involves three operating stations (Fig. 5-18). The first is the injection station where the parisons are molded. The second is the blowing station where the parisons are transferred to blow molds for the introduction of high-pressure air. The third is the ejection station where finished containers are stripped from the core rods. One variation from the norm is the addition of a fourth station for detection of nonejected containers, for free-blowing parisons, or for in-line decoration.

A complete cycle involves injection molding of the parison, indexing of the turntable to the blowing station where air is injected through spring-loaded core rods, and finally indexing to the eject station where containers are automatically stripped off onto a conveyor or into a carton. The line is highly productive, requiring little operator attention.

The three station machines with horizontal platen indexing comprise one group, and another group, operating on a rotating-arm principle, is available (Fig. 5-19). In this system, the parison is molded at a twelve o'clock position. The arm then rotates 180° after the mold opens and is captured by the blowing mold. At the same time, the second arm rotates back to the injection station, a new preform is injected, and a container is blown at the 6 o'clock position. Another method of transferring core rods involves a shuttling-back-and-forth process.

Some injection blow molding systems are designed as blow-station packages that can be attached to certain standard injection molding machines. The majority, however, are sold as complete injection blow molding systems. The extruder can be mounted either vertically or horizontally, and, as discussed earlier, they can be either reciprocating or nonreciprocating.

Dip Mandrel System

Although not a pure injection molding process, dip mandrel blow molding is closer to injection molding than to extrusion. Simply stated, the process involves dipping mandrels into a melt accumulator, then transferring the resin-coated mandrels to a blow molding station for container formation (Fig. 5-20). While not new, the process is not as popular in the United States as in other parts of the world.

In dip mandrel blow molding, the extruder continuously feeds two open-ended reservoirs, into which two mandrels descend to be coated and ascend to be transferred to blowing stations. The mandrels descend into the reservoir, and a neck-forming block seals the top of the reservoir. Electronically controlled plungers retract as the mandrels drop. The sealing of the top opening

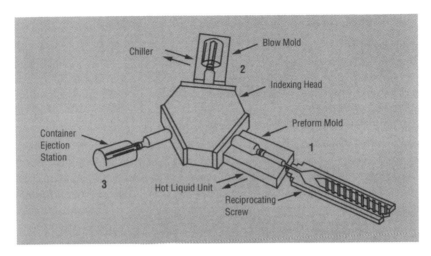

Fig. 5-18 Injection blow molding. 1, Injection station. 2, Blowing station. 3, Ejection station. Courtesy of Quantum, USI Division

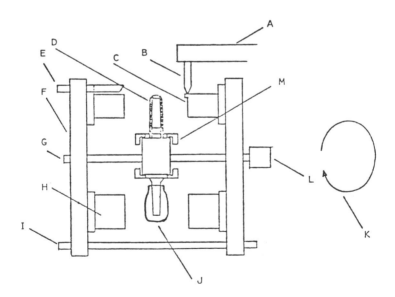

Fig. 5-19 Rotating-arm injection blow molder. A, Extruder. B, Nozzle. C, Preform mold. D, Preform. E, Tie rod. F, Platen. G, Rotation tie rod. H, Blow mold. I, Tie rod. J, Finished container. K, Direction of rotation. L, Rotary actuator. M, Neck ring slides

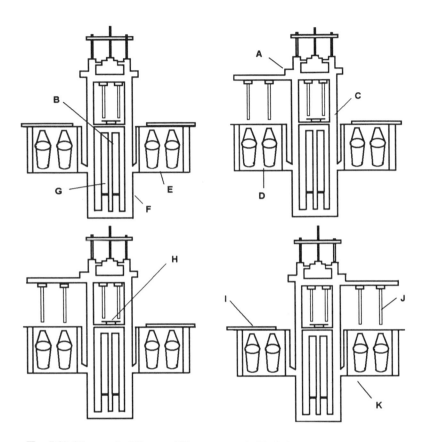

Fig. 5-20 Dip mandrel blow molding process. A, Neck forming. B and C, Dipping mandrels. D, Blow molds. E, Blow cycle. F, Melt accumulator. G, Plungers. H, Cutting tool. I, Blow cycle. J, Parisons. K, Container eject

allows the plungers to inject melt into the neck-forming tools, which allows for precise control of thread tolerances. The mandrels and plungers then move, extruding a tube onto the mandrels. The mandrels shift to one of the two blowing stations, and the other station ejects containers and accepts another pair of parisons. The cycle is totally continuous, in that something is always occurring and there is no interruption of the blowing process.

An important feature of this system is that the weight of the parison, and thus of the container, can be varied by varying the speed of the plunger.

Slowing the plunger thins the parison, and speeding the plunger thickens the preform.

Container Types

Injection blow molding has had the reputation of being limited to small containers. It is true that very small containers are more profitably made by injection blow molding than by any other system. This does not, however, restrict the process to mini-containers. In the drug and cosmetic fields, where 1 to 8 oz containers are widely used, injection blow molding has achieved a firm position because:

- Fast processing and multiple cavities increase productivity.
- The close dimensional tolerances achieved are necessary to these industries.
- Some FDA-approved resins used in the drug industry are not as easily processed by the blow molding systems.
- The high degree of decoration required can be achieved through readily available, totally automatic decorating systems.

Recent advancements in injection blow molding machinery have allowed a significant trend toward production of larger containers. One-, two-, and three-liter bottles are commonly produced, and with consumer demand high for larger containers, production of one- to five-gallon containers will soon be common. Penetration into the extremely large beverage container market has been the guiding force behind this trend, with resin suppliers continuing to improve plastic formulations to meet customer requirements.

The industrial market has also shown outstanding promise for injection blow molding. Polycarbonate light globes for street lighting present one interesting opportunity. Automotive parts that could not otherwise be made are being tested. The noncontainer market appears to be a major outlet for future injection blow molded parts.

Biaxial Orientation

Following breakthroughs in resin technology for the carbonated beverage industry, biaxially oriented containers have surfaced as viable solutions to difficult liquids-packaging applications for food, beverage, cosmetic, and household products.

Simply defined, biaxial orientation (also known as stretch blow molding) involves stretching a parison or preform lengthwise mechanically and radially through the blowing process. The stretching must be done after the parison is solidified, with a very narrow temperature range that varies from resin to resin. The stretching from 20 to 200% in each direction orients the resin molecules, and the end result is a significant increase in container strength (and with some resins, an increase in container clarity as well). These properties are achieved at container weight reductions of 25 to 33%.

Systems. Two basic procedures are currently available for achieving biaxial orientation: in-line systems and two-stage systems.

In-line systems have a number of variations. The key ingredient in stretch blow molding is the ability to bring the parison or preform to proper orientation temperature, so the major innovation in the in-line system is the addition of a sophisticated conditioning station to accomplish this critical temperature control. In this type of system, the parison or preform is brought to the proper orientation temperature, stretched, and blown without cycle interruption. Some machines of this type also afford parison programming for greater wall thickness control.

Two-stage systems, often called reheat systems, differ from the in-line process in that preforms are usually cooled to room temperature, then reheated later to proper orientation temperature before being stretch blown into containers. These systems generally require two machines, one for preform molding and the second for stretch blowing the product.

On the other hand, preforms can be purchased elsewhere or manufactured in one location and shipped to satellite molding plants. The reheat system may have drawbacks with certain heat-sensitive resins that require minimum heat history. Preforms are generally loaded in bulk and automatically fed into transfer stations that move them through the conditioning oven and into the blow station. The parison or preform resembles a threaded test tube.

Methods. The methods used to form the parisons and stretch blow the containers can vary significantly, and a number of combinations are used on the machinery available to date. Preforms can be extruded, injection molded, or blow molded. The orientation or stretching can be accomplished by mechanically pulling the preform or by using a telescoping mandrel that extends from the blow pin down inside the preheat form (Fig. 5-21). The telescoping mandrel accomplishes the lengthwise stretching and is used on both parison systems and preform systems. In all cases the radial stretching takes place during the blow operation of the cycle. Mechanical stretching lengthwise is also accomplished by clamping and sealing off one end of the parison, then pulling it downward (upward in some machines).

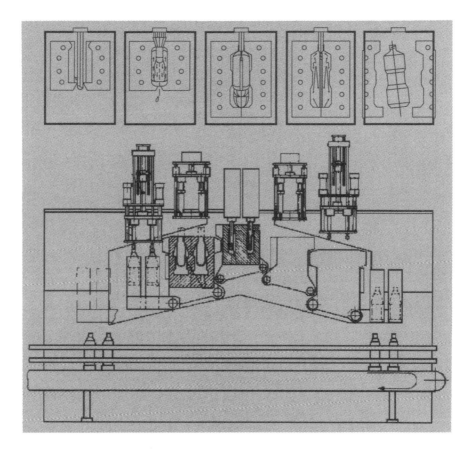

Fig. 5-21 Biaxial orientation (stretch blow molding)

Parisons or preforms can be either extruded, injection molded, or, in one case, compression molded. Injection blow molding machines have been developed that use a fourth station as the parison conditioning station for orientation purposes.

Advantages. The major advantages of using oriented containers are:

- Increased strength
- Clarity
- Environmental stress crack resistance

- Luxurious feel
- Weight reduction

The cost savings are the real advantages, although aesthetic value also results in increased product sales, thus increasing container consumption.

In the enormous carbonated beverage container market, the use of oriented containers is essential to solving the gas permeability problem these beverages present. Standard blown containers would require extremely thick (and costly) walls to achieve the same properties.

Machinery, resins, and technology in the oriented container field are advancing so rapidly that it is now feasible for a small custom blow molder to enter this field with reasonable confidence. The typical oriented containers have been two-liter pop bottles and containers for window cleaners, dishwashing detergents, and orange juice, but advances are allowing other products, such as liquors, pharmaceuticals, and chemicals, to use oriented containers.

Molds

Efficient forming and cooling of a container is totally dependent on molds of proper design. Figure 5-22 shows a typical extrusion blow mold and its component parts. The dark area that forms the partial outline of the container is known as the pinchoff area. The pinchoff area is beveled, sharpened metal that squeezes the parison into a narrow membrane-like form and seals it at the top and bottom contours of the container. The excess plastic can then be easily trimmed off with automatic or manual equipment.

In the opposite top and bottom corners of the mold, guide pins are inserted that match female bushings inserted into the corresponding corners of the other mold half. These guide pins and bushings facilitate mold alignment during the installation of the mold into the blow molder, and they keep the mold halves in line during operation of the machine. Misalignment of the mold halves after extended operation can often be traced to worn pins and bushings or to worn platen guides on the machine itself.

Flash Pockets in Molds. In the flashed neck method of extrusion blow molding, flash pockets for the proper depth are incorporated into the molds to sufficiently cool the excess flash, which is usually considerably thicker than any portion of the actual container. Flash relief areas are then machined to a depth sufficient to squeeze a double thickness of parison with enough force to allow solid contact between the flash and the cold mold material. Too shallow a flash relief area will not allow the mold halves to completely close, and too

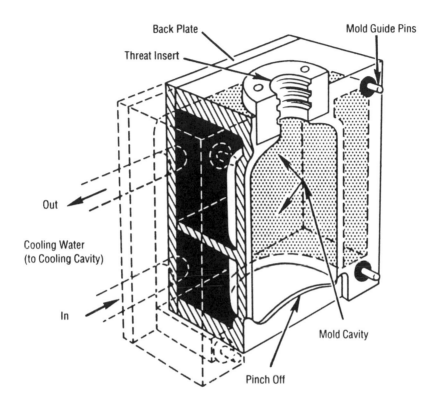

Back Plate

Mold Guide Pins

Threat Insert

Out

Cooling Water
(to Cooling Cavity)

In

Mold Cavity

Pinch Off

Fig. 5-22 Schematic of an extrusion blow mold. Courtesy of Quantum, USI Division

deep a relief area will result in insufficiently cooled or wet-looking flash that is difficult or impossible to trim. The same relief path applies to both neck and tail flash. The depth of the flash pockets is determined by the container weight, and thus by the thickness of the parison.

Once molds are machined, only a limited weight range for that container is possible. A standard 50 g quart, for example, could run between 40 and 60 g, but to attempt to run the container outside this weight range would be almost impossible without mold modifications.

Another flash-cooling method, used especially on extra-light containers, is called blown flash. In this system, air is blown into the neck flash, which in turn is blown into specially machined pockets in the neck block. The flash is cooled by the air and by contact with the cold neck block.

Mold Venting. When molds close on the parison or preform, air is always trapped between the outside of the parison wall and the inside of the mold wall. This atmospheric air must be vented in order to completely inflate the container against the mold walls. Venting is accomplished in two ways:

- Most of the trapped air escapes along the parting line of the two mold halves that have been machined to allow air passage. In many cases, porous inserts are added along the parting line to help the venting process.

- Vents are machined or drilled into the mold body. The most common position for one of these vents is at the farthest point away from the center of the container, which is usually on a shoulder. Small plugs of porous material are also used in this location. The air escapes through the pores in the plugs, through small holes drilled into the back of the mold to meet these plugs, and through a clearance slot running from the top to the bottom of these mold backs (Fig. 5-23). Vents from the next insert and from the tail block section of the mold can also be tied into this vent channel.

Insufficient venting can result in weak spots in the container. In severe cases, it can result in rough container surfaces or even distorted containers, because air trapped between the plastic and the mold will not allow proper contact between the plastic and the mold wall, resulting in insufficient cooling of the container.

Mold Cooling. All blow molds must be cooled by a continuous flow of cold liquid during the automatic cycle to remove the heat from the plastic melt. Cooling passages are drilled into the mold block in such a manner as to form a loop for the cold liquid to flow through. Most often, coolant enters the mold from the bottom, flows up one side, across the top, down the other side, and out. The source of the coolant is usually a liquid chiller equipped with a circulating pump. The system is closed: after the coolant has removed heat from the molds, it returns to the chiller to be cooled again. The coolant used is generally a 50% water/ 50% ethylene glycol solution. Some molders use coolant chilled by a cooling tower, while a few others use well or city water. Temperature control with either of these systems is difficult, and in the case of city water, expensive.

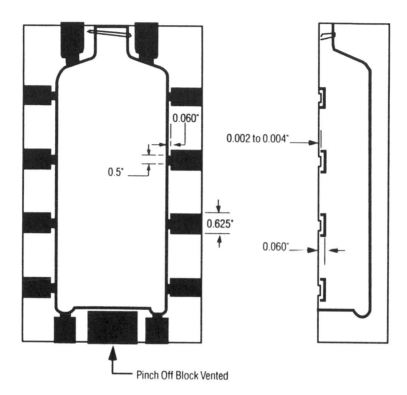

Fig. 5-23 Venting of an extrusion blow mold. Courtesy of Quantum, USI Division

In injection blow molding, the blowing molds are also cooled by chilled water. However, the preform molds are "cooled" by hot water, or hot oil in many cases. Hot water or oil is still at a lower temperature than the plastic melt, so it cools the parison just enough that the parison can be effectively blown into a container. Cold water to the preform molds would cool the parison too much and make blowing impossible. On the other hand, if there was no cooling, the parison might be too hot to be transferred to the blowing station. Cooling takes place primarily in the neck area of the parison.

Mold Temperature. The effectiveness of mold cooling depends on the temperature of the coolant and the velocity at which it flows through the molds. A common misconception is that the lower the coolant temperature, the better the containers will cool and the faster the cycle time. With the fast cycle times being run, this is not entirely accurate. The variables in this theory are the temperature of the melt, the thickness of the parison, the size and location of mold cooling passages, and the velocity of the coolant. The important thing to identify is the temperature change of the coolant as it enters and exits the mold. A change from 5 to 10 °F is desirable, and this can usually be accomplished with an inlet coolant temperature of 50 °F, providing the melt temperature is not too high.

Mold Material. Blow molds can be manufactured from a number of materials:

- Forged aircraft aluminum is probably the most common. It is reasonably priced, lightweight, and easy to machine, and it has good heat transfer properties. However, it is reasonably soft, making pinchoff areas vulnerable to damage.
- Beryllium-copper is strong and possesses excellent heat transfer properties. However, it is expensive and heavy, making mold changes more difficult.
- Steel and Kirksite each have some merit.

The recent trend in mold making has been to combine two of the above materials. For example, an aluminum body with beryllium-copper pinchoff inserts combines the advantages of each material. For some resins, such as PVC, steel and beryllium-copper are the only feasible mold materials. Steel needs to be chrome plated to provide a corrosion-resistant barrier, because PVC tends to release hydrochloric acid from the polymer melt, which can cause pitting in the unprotected mold cavity.

Tooling

In blow molding, the term *tooling* refers to the components used in the formation of parisons or preforms. The tooling for extrusion blow molding and the tooling for injection blow molding differ significantly.

Tooling for Extrusion Blow Molding. Tooling for extrusion blow molding consists of dies, mandrels, and head components. The number of heads in a system can vary from one to twelve, and each requires the same components, precisely reproduced. Figure 5-24 shows a typical extrusion head and its components.

In single-head operation, the important factors are the ability to control parison thickness, adjust for straight-falling parisons, and achieve some shift-

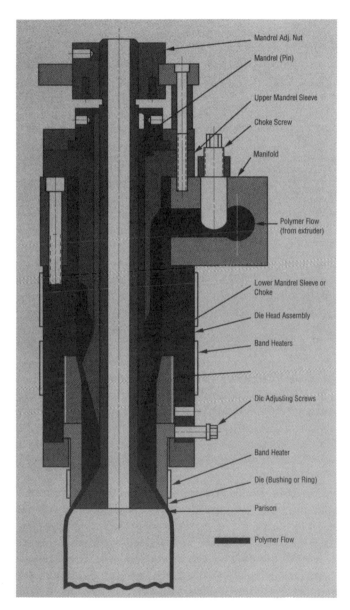

Fig. 5-24 Typical extrusion die head. Courtesy of Quantum, USI Division

ing of melted material to desired points. In multiple-head extrusion, the ability to balance material flow to all heads is important in addition to the above requirements. The thickness of the parison and the weight of the product are controlled by moving the mandrel up and down to open or close the die gap. A straight-falling parison is controlled by centering the die on the mandrel. In like manner, material can be shifted to points around the parison circumference to some degree by off-centering the die on the mandrel. Flow balance to various heads must be accomplished with some type of choke or flow restructure on each head.

Tooling for Injection Blow Molding. Injection blow molding requires a completely different type of tooling (Fig. 5-25). Parisons require molds and core rods that are similar to those used in standard injection molds. The components needed are:

- *A two-piece mold body,* machined to the desired shape of the preform
- *Core rods* that have spring-loaded tips, to allow opening of the air passage during blowing and closing during injection
- *Neck rings* that form the threaded portion of the parison
- *Injection nozzles* that allow for flow of melt from the extruder into the parison mold

The components are attached to die sets for mounting in the machine. Very close tolerances are necessary on both the parison mold bodies and the die set fit. The fit of the core rods into the parison cavities is also critical. The shape, size, and weight of the parison, as well as the material distribution within it, are all accomplished with this tooling, in addition to the molding of precision threads on the container neck. Unfortunately, injection blow molding tooling is expensive because of the precision needed and because new tooling is needed for each new container.

Tooling Designs. The most common changes in tooling design are done in the die and mandrel area, for these two components control the size and shape of the parison to be extruded. Internal head components are generally supplied with the machine and do not need to be changed from product to product. The main functions of these components are to direct the flow of the melt through the heads and to accomplish uniform welding of the melt.

Welding is the term used for the joining of the melt as it flows around the mandrel sleeve to form a cylinder. The welding is accomplished through back pressure exerted in the head as the melt tries to pass through the narrow gap between the mandrel and the centering ring, and it is necessary to achieving a

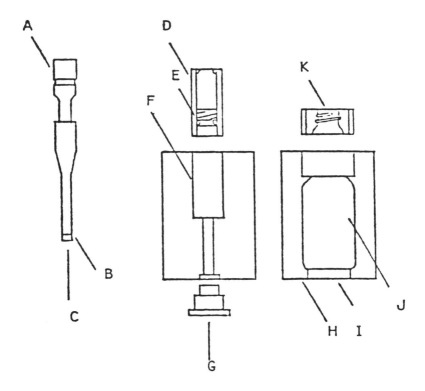

Fig. 5-25 Injection blow molding. A, Core rod. B, Blow air entry point. C, Core rod tip. D, Preform neck insert. E, Preform neck. F, Neck ring pocket. G, Sprue bushing/ nozzle. H, Blow mold. I, Blow mold bottom plug. J, Container cavity. K, Blow mold neck insert

uniform, seamless tube. The back pressure necessary for proper welding is controlled by proper positioning of the mandrel sleeve in relation to the centering ring.

Die and mandrel sizes and shapes are totally dependent on the size, shape, and complexity of the container to be blown. Serious parison production problems can often be traced to improperly matched dies and mandrels. There are three basic variables to be taken into account in die and mandrel designs:

- *The diameter of the mandrel* is the most important in relation to the parison size.
- *The exit angle of the mandrel* determines flare.
- *The land length of the angled surface* also determines parison quality.

Changing any one or more of these variables significantly changes parison quality, and usually only a slight modification is necessary to produce the change.

Equipment Operation

Proper equipment operation is important not only to equipment life and machine efficiency, but also to the quality of the containers that will be produced and to reducing the severity and frequency of maintenance problems. Improper equipment operation often results in poor parison quality, which is the primary thing machine operators should be consistently striving to avoid.

Start-ups

The frequency of start-ups varies greatly from plant to plant, depending on the number of shifts run per day and the number of product changeovers. A typical start-up involves bringing a cold machine up to operating temperature, stabilizing the flow of the melt through the machine, and adjusting for balanced, uniform parisons.

Operating temperatures vary with the resin used and to some degree with the process used. Resin manufacturers normally recommend a temperature range in which to work. Attempting to operate at the lower end of the range is generally advantageous.

Purging, or stabilizing the flow of the melt, involves extruding or injecting a number of shots onto a clean surface. Observation of these shots indicates whether the melt is properly mixed and ready for molding, whether there is too much or too little heat in the melt, and whether adjustments need to be made.

Adjustments

Adjustments to the blow molding machine are usually straightforward, but the frequency of adjustments is greater than one would expect, especially with extrusion blow molders. Some operating adjustments are routine: the heads, extruder speed, cycle time, and clamp on extrusion blow molders; and the

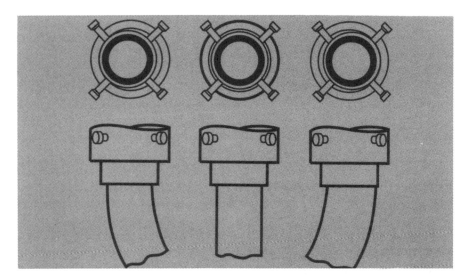

Fig. 5-26 Adjustment of die bolts to prevent parison curving. Courtesy of Quantum, USI Division

injection pressure, cycle time, and clamp on injection blow molders. Injection blow molders are much less operator-intensive and usually require little attention once in automatic operation.

Head adjustments require the most consistent attention on extrusion blow molders, because the head or heads determine the shape, size, and quality of parisons. The thickness of the parison, and thus the weight of the container, can vary from cycle to cycle.

Die and mandrel adjustments are probably the most common of all normal adjustments. With diverging tooling, raising the mandrel closes the die gap and thins the parison. Lowering the mandrel has the opposite effect. With converging tooling, the procedure is reversed. Centering the die on the mandrel controls the straight or crooked fall of the parison (Fig. 5-26). With the die centered, parison wall thickness is equal all around and the parison falls straight. Off-centering the die allows for the shifting of wall distribution, but it also results in a hooking parison. Weak spots in a container can often be remedied by centering the die.

Parison length has a significant effect on system efficiency, particularly on containers that are ejected with flash intact. Transfer of the untrimmed container from the blow molder to the trimmer must be smooth and quick. Too short a parison may result in a partially blown container and poor transfer, while if the parison is too long, excess tail flash may be ejected in molten form, which causes poor transfer and causes containers to stick together.

Parison length can be controlled by adjusting:

- *Extrusion speed:* Too fast a speed results in excessive parison lengths, and vice versa.
- *Weight,* particularly on multiple-weight machines. Unbalanced parison weights result in unbalanced parison lengths.
- *Cycle time,* which in essence is tied into extrusion speed. A balance between the two is essential to proper parison production.

Mold clamp adjustment is possibly the most critical adjustment on a blow molding machine, yet it is the most abused. Excessive clamp force can ruin a set of molds quicker than anything short of taking a hammer to the molds. It has been a general practice to overclose molds between 0.010 and 0.060 in. to achieve a good lock-up. This means that once mold faces just touch in the closed position, an additional length of travel within the above range is added to effect a tight pinchoff.

With proper alignment, parallel surfaces, and properly machined pinchoffs, the necessity of overclosing diminishes and the life expectancy of mold pinchoff increases. In many cases, clamp adjustment is improperly used as a cure-all for a variety of problems. Although a temporary solution may be achieved, the long-term result is reduced mold life.

Shot Adjustments. The amount of melt delivered in any sample shot can be controlled and adjusted. In extrusion blow molding, the length of travel of the screw and the ram cylinder are the primary controls. In injection blow molding, the pressure with which the melt is injected into the preform molds has a significant effect on the quality of containers produced. Excessive pressure will result in flash around the parting lines, while inadequate pressure will result in unfilled mold cavities, often referred to as short shots.

Temperature control over the molding equipment is critical to controlling the final melt temperature. This process is usually a heat-in/heat-out procedure, but achieving this supposedly simple process is often not an easy task. Because frictional heat in the barrel is the major contributor, efficient cooling of the barrel is the key counterbalance. One simple method of increasing this cooling efficiency is to minimize the amount of external heat applied by keeping the temperature controller setting as low as possible. Remember that excessive external heat also contributes to degraded polymer.

Shutdowns

The ability to start up with ease can often be influenced by the procedure used to shut down. In many cases, after blowing the last cycle of the day, shutting down simply involves turning off the power to the blow molder. However, certain heat-sensitive resins may require easing the temperature profile down by continuing to extrude melt while decreasing temperatures. This is true when high barrel temperatures are used. The purpose is to minimize oxidation or crosslinking of the polymer molecules, which results in degrading of the polymer and "dirty" melt on the next start-up. Complete purging of heat-sensitive resins at each shutdown is also common practice.

In some cases, turning temperatures back on the barrel is used as a shutdown procedure. Not only does this contribute greatly to the degradation of the polymer, it can also be a serious safety hazard should the temperature controller fail. Further, this procedure is a waste of valuable energy, so every effort should be made to turn off equipment during extended shutdown periods.

6

Injection Molding

Injection molding is singularly the most popular of all the plastics processes. Unlike extrusion or blow molding, it allows the plastics processor to produce a plastic part with three-dimensional characteristics. This permits very intricate designs and high production rates.

Injection molding dates back to the early twentieth century. The first simple machines consisted of no more than a manual arbor press used to force a plunger through a heated barrel that contained the plastic material. This arbor-plunger system then forced the plastic melt into a closed mold that produced the first accurate three-dimensional plastic parts. As plastic materials developed, so did the means of processing them. Injection molding machines and their molds developed in both size and complexity in the 1950s and have been significantly refined since.

The nature of the injection molding process and its ability to produce complex shapes at a fast rate led part designers and manufacturers to consider plastic parts an alternative to conventional materials, such as wood, metal, ceramic, or glass. The simultaneous development of plastic materials and injection molding systems was the cornerstone of today's enormous plastics manufacturing industry.

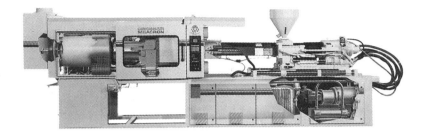

Fig. 6-1 Injection molding machine. Courtesy of Cincinnati Milacron

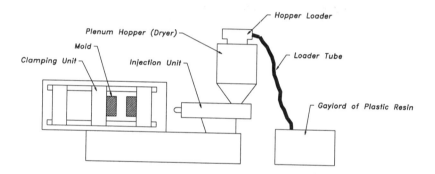

Fig. 6-2 Typical configuration of a production molding machine

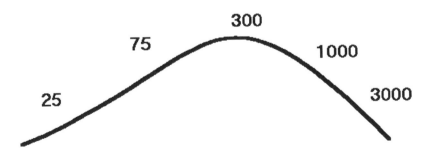

Fig. 6-3 Distribution of injection molding machines produced today, based on clamp size as shown in tons

The Injection Molding Machine

Today, the injection molding machine is defined and characterized by its clamp size and its injection capacity. The term *clamp size* means the force, measured in tons, available to hold the mold closed during the high-pressure injection of the plastic melt. Clamping force may range from 5 tons for a tabletop molding machine to 10,000 tons for a mammoth machine the size of a large house. The term *injection capacity* relates to the injection unit of the molding machine, which melts the plastic and forces the plastic melt into the closed mold. The pressure under which the plastic is injected is extremely high, ranging from 500 to 30,000 psi. The potential part weight, or shot size, of an injection molding machine can range from a few grams to several hundred pounds.

Figure 6-1 is a photograph of a modern injection molding machine. Figure 6-2 shows a typical configuration for a production molding machine. The distribution of the machine sizes produced today, based on clamp size measured in tons, is depicted in Fig. 6-3. The clamp size of the average injection molding machine is 300 tons. To estimate the cost of such a machine, a rule of thumb is that machines with clamp sizes between 50 and 400 tons cost $1000/ ton. Thus, a top-quality 300 ton press with modern control features will cost approximately $300,000. Needless to say, injection molding is a capital-intensive manufacturing process.

Over the past 20 years the top manufacturers of injection molding machines have shifted in a pattern very similar to that of manufacturers of other heavy equipment. The United States, Austria, Canada, Germany, Japan, and Korea are now the key players in the manufacture of injection molding machines.

The Injection Unit

The injection unit of an injection molding machine is where the plastic pellets are melted and the plastic melt is forced into the mold. The entire injection unit rests on a sled that guides the unit forward to contact the mold. Alternatively, the unit can be moved rearward to allow the processor to purge the plastic out of the barrel. The injection unit is very similar to an extruder in design and construction, particularly in that it has an alloy barrel and a flighted screw. However, unlike an extruder, the injection screw is capable of moving back and forth in a reciprocating motion. This has led to the name *reciprocating injection molding machine.*

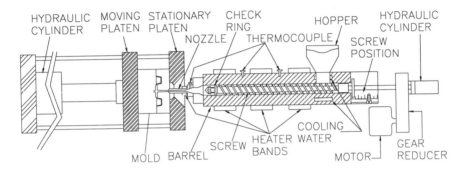

Fig. 6-4 Detailed view of an injection molding machine

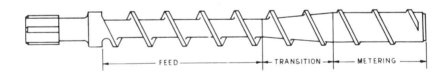

Fig. 6-5 Injection screw. Courtesy of Spirex

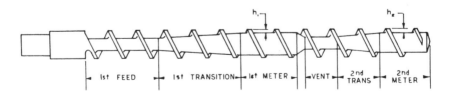

Fig. 6-6 Two-stage screw. Courtesy of Spirex

Figure 6-4 gives a closer look at the injection unit. The barrel allows plastic pellets to enter through a water-cooled throat. As the screw rotates, the plastic pellets are augured forward. As in the extrusion process, most of the heat used to melt the plastic pellets evolves from the friction between the plastic pellets and between the barrel and screw, not from the electric heater bands that surround the barrel. Instead, the electric heater bands provide the initial thermal energy required to start the injection molding process, and they provide a thermal barrier to prevent the heat generated within the barrel from escaping.

Injection Screw. The injection screw must perform several functions during the molding cycle. In addition to conveying and melting the plastic, the screw must mete out the correct amount of plastic in front of the screw, and it must inject this plastic melt into the mold.

Figure 6-5 illustrates a typical injection screw, which has features similar to those of an extrusion screw. The size of the screw root gradually increases, which increases the shearing and mixing action and melts the plastic pellets. The screw is categorized by its L/ D ratio (ratio of length to diameter), typically 20:1 to 24:1, and by its compression ratio (ratio of the depth of the flight in the feed section to the depth of the flight in the metering section), typically 2.5:1 to 3:1.

The screw also has distinct areas, each with its own function. The material that enters the barrel is considered to be in the feed section of the screw. The feed section will have predominantly unmelted plastic pellets. As the pellets are augured forward toward the middle zone of the screw, they begin to melt. This area is called the transition section, because there will be both soft pellets and plastic melt. As the plastic becomes completely melted, it moves toward the forward area of the screw, the metering section, which must accurately deposit it in front of the screw so that it can be injected into the mold.

The two-stage vented screw is a variation of the typical injection screw (Fig. 6-6). It has a primary transition and metering section midway down its length. At the end of the first metering section, there is a significant narrowing of the screw root, which lowers the pressure of the plastic melt and allows the barrel to be vented. The barrel vent is essentially a hole in the barrel wall. If a standard injection screw was used, the plastic melt would be forced out of such a hole, rendering the processing system useless. With the two-stage screw, however, the lower material pressure in the area of the barrel vent reduces the risk that plastic melt will exit. The vent does allow moisture and other volatiles to be removed, and it may eliminate the need for some ancillary equipment, such as material dryers, because any moisture present in the plastic is converted to a gas by the time the plastic melt reaches the vent. After the vent, the screw root again begins to increase in diameter, creating a second transition zone and a second metering zone, after which the screw performs like a typical injection screw.

A *check valve* is attached to the discharge end of thermoplastic injection molding screws. To understand its function, it is necessary to understand the flow of plastic melt in front of the screw. When the material in front of the screw reaches sufficient pressure, it forces the still-turning screw rearward. The screw continues to turn, but it is forced rearward until it reaches a predetermined setpoint, called the shot size, which determines how much plastic melt will be allowed to advance in front of the screw. At this set point, the screw stops turning.

The next step in the process is for the screw to move forward at a controlled rate of speed and pressure. As the screw moves forward it does not turn, and its plunger-like action forces the plastic melt into the mold. During this forward plunger action, it is critical that the plastic melt in front of the screw not be allowed to flow rearward, back over the screw. To prevent this from happening, a check valve seals off, preventing the plastic melt from moving forward or rearward.

There are two basic types of check valves. The sliding ring check valve (Fig. 6-7) works by having a precision-machined ring slide forward as the plastic melt applies pressure during screw rotation. Plastic melt flows under the ring and in front of the screw. During injection, the ring slides rearward, sealing off against a seat. This prevents plastic melt from being forced back into the metering section of the screw during the injection phase of the molding cycle.

The ball check valve (Fig. 6-8) works in a similar manner, except that a ball is forced against a seat. Plastics molders debate the pros and cons of each check valve system. However, most agree that the sliding ring check valve is more streamlined, easier to clean, and easier to maintain, while the ball check valve offers a more positive shutoff.

The check valve assembly is attached to the discharge end of the screw with a left-handed thread, so that it won't unthread itself when the screw rotates during the molding cycle. Technicians who understand this fact may save themselves a lot of frustration the first time they attempt to remove one.

While most injection molding machines use a check valve, there are some situations where it may be eliminated. Examples are thermoset injection molding, discussed later in this chapter, and rigid PVC molding, in which high heat and high shear forces are generated when material passes around the valve components.

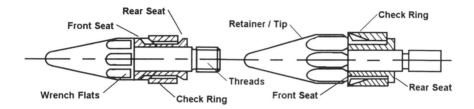

Fig. 6-7 Sliding ring check valve. Courtesy of Spirex

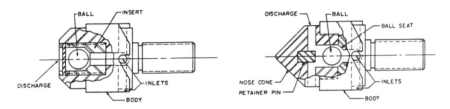

Fig. 6-8 Ball check valve. Courtesy of Spirex

Screw Diameter. One of the most overlooked aspects of the injection screw is the area reduction associated with the diameter of the screw and the tip of the screw. Regardless of whether a check valve is used, the diameter of the screw and screw tip is significantly less than the area of the rear hydraulic cylinder where pressure is applied. Depending on the screw diameter and tip design, most screws have approximately a 10-to-1 reduction in area between the hydraulic cylinder and the screw diameter. This area reduction increases the effective injection pressure to a level 10 times that of the hydraulic pressure applied to the rear of the screw.

The technician should be aware of this basic concept and know whether the injection pressure gauge on the molding machine reads the hydraulic pressure on the screw or the actual injection pressure on the plastic. Many injection molding machine manufacturers attach a pressure conversion graph to help the processor determine the correct plastic injection pressure.

Barrel Capacity. Residence time is the time it takes a plastic pellet to journey from the throat, where it enters the barrel of the injection unit, to the nozzle, where it exits the barrel. The calculation of residence time is important to understand, because if the plastic spends too much time in the heated barrel it may degrade, resulting in a change of appearance, a reduction in properties, or both.

Injection molding machine manufacturers usually rate the capacity of an injection unit in terms of ounces of general-purpose polystyrene (GPPS). For example, a 10 oz injection unit can inject up to 10 oz of GPPS. It is unlikely that any processor would use 100% of the injection unit capacity; most processing guides suggest that 70 to 90% of the capacity be used for the shot size of a particular mold. Less than 70% will allow the plastic to reside too long in the injection unit, which may result in thermal degradation of the plastic. Using over 90% of the capacity may leave insufficient material for process adjustments and for cushion. (Cushion is defined in the section "The Injection Molding Process" in this chapter.)

Residence Time. To determine the residence time of plastic within the barrel, the processor needs to know the weight and volume of the plastic held within the barrel when it is full. This information is not easily determined, because a processor cannot readily completely evacuate the plastic melt from the barrel. Therefore, the molding machine manufacturer should be consulted.

Example: Calculating Residence Time. A plastics processor is using a 10 oz (GPPS) capacity molding machine to process glass-filled nylon. The mold requires a shot size of 8 oz. The molding cycle time is 30 sec. What is the residence time of the plastic in the barrel?

- *Improper Method:* Machine capacity is 10 oz, or 25% greater than the shot size. Each shot is 8 oz, or 80% of capacity. Therefore, residence time is 125% of the 30 sec cycle time, or 37.5 sec. *This is totally incorrect!*

- *Proper Method:* Contact the machine manufacturer to obtain the barrel capacity (plastic material contained in a full injection unit). Let's say it is 40 oz of GPPS. We must then correct the barrel capacity for the glass-filled nylon. The specific gravity of GPPS is about 1.04, and the specific gravity of 33% glass-filled nylon is 1.34. The difference in specific gravity means that the barrel will be able to hold (1.34 ÷ 1.04) times as much glass-filled nylon as GPPS, or 1.29 times as much weight. Therefore the corrected barrel capacity for the nylon is 1.29 × 40 oz, or 51.6 oz. Every 30 sec, 8 oz of nylon is injected into the mold. At this rate, it will take 51.6 oz ÷ 8 oz/ cycle, or 6.45 cycles, to use all the plastic in the barrel. Thus the residence time is 6.45 cycles × 30 sec/ cycle, or 193.5 sec (3.23 min). If a mold half the size was placed in the molding machine, the plastic material would reside twice as long, or nearly $6\frac{1}{2}$ min, within the barrel.

Residence time is often overlooked by processors. The major definition of machine size is clamp tonnage, and too many times a mold is set up in a molding machine simply because it physically fits within the clamping capacity.

Heat Control. Thermoplastic injection units are heated by electric heater bands that surround the barrel. The physical size and power output of these bands is specifically designed to provide optimum control and heat density. The major portion of the thermal energy is provided by the frictional energy of the plastic within the barrel, so the heater bands may be thought of as a control valve that adds heat or allows heat to escape the barrel, depending on the process requirements.

Over the long life of an injection molding machine, some of the heater bands will require replacement. Too often, the selection of a replacement is based not on what the machine manufacturer has defined, but on whether there is a heater band available that will physically fit onto the barrel. (Depending on the size of the injection unit, heater bands can be quite expensive, and therefore most processors keep only a minimum inventory on hand.) Replacing a heater band with one of similar physical size, but different power output, may resolve the immediate maintenance problem, but it may cause significant processing and process control problems later. Heaters on injection units should be inspected periodically, and the performance and rating characteristics should be verified.

Barrel heaters are controlled by thermocouples or RTDs, which sense the barrel temperature, and by proportional temperature controllers, which maintain the correct energy output. However, the heater band for the nozzle (discussed below) is often less sophisticated in its control. In addition, the nozzle is out of the screw area, so the thermal energy required to keep the plastic melted there comes only from the nozzle heater band. As a result, the nozzle heater band is one of the most difficult to control and easiest to damage.

The Nozzle. The nozzle is the extension on the discharge end of the barrel that allows the plastic melt to exit the barrel and enter the mold through the sprue bushing. It may be one piece in design, or it may have a removable tip to facilitate cleaning and materials changeovers. The injection unit is held against the sprue bushing by the action of a hydraulic cylinder that maintains pressure on the sled that supports the injection unit. The alignment of the nozzle is controlled by the alignment of the injection unit and the sled on which it rides.

The nozzle must be properly matched and aligned to the sprue bushing. Otherwise, plastic will leak from between the nozzle and sprue, and the process will deteriorate. The radius of the nozzle tip and the hole through which the plastic melt is injected must conform to the radius and hole of the sprue bushing. Most often these radii are either $\frac{3}{4}$ or $\frac{1}{2}$ in. If there is a problem with matching the nozzle and sprue, plastic will not flow properly into the mold.

Figure 6-9 illustrates four nozzle-to-sprue alignment possibilities:

- *Correct alignment:* The sprue radius matches the nozzle radius, and the orifice of the nozzle is smaller than the orifice of the sprue (Fig. 6-9a).
- *Misalignment of the nozzle,* regardless of the radii and orifice match, will result in plastic leakage (Fig. 6-9b). This problem is most often corrected by adjusting the injection unit sled.
- *Improper matching of the nozzle and sprue radii* will cause leakage (Fig. 6-9c).
- *An oversized nozzle orifice,* larger than the orifice of the sprue bushing, will cause the sprue to stick within the sprue bushing at the point of the nozzle (Fig. 6-9d).

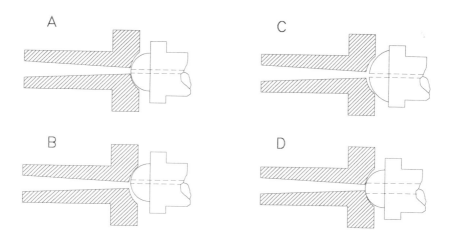

Fig. 6-9 Four nozzle-to-sprue alignment possibilities. (a) Correct alignment. (b) Misalignment of the nozzle. (c) Improper matching of the nozzle and sprue radii. (d) Oversized nozzle orifice

Nozzle Tip Configurations. There are many different types of nozzle tip configurations (Fig. 6-10). A standard nozzle tip usually has a straight or forward taper (Fig. 6-10a). These tips work well for most amorphous thermoplastics, for glass-filled plastics, and for some semicrystalline materials.

A reverse tapered nozzle (Fig. 6-10b) provides a slightly widening orifice to facilitate removal of solidified plastic material from the tip. This configuration is suggested for some semicrystalline plastic materials, such as nylon, that have a tendency to "freeze off" (solidify easily).

The cause of freezeoff is that the nozzle tip loses heat to the colder sprue bushing in the mold. When a semicrystalline plastic such as nylon is being molded, too often the molding cycle is constantly interrupted because the plastic freezes off and forms a plug that is not displaced when the next shot is injected. It may not occur to the technician to correct nozzle temperature or inspect the nozzle heater bands. If a reverse tapered nozzle is not available, the technician often resorts to placing a piece of cardboard between the nozzle tip and the sprue bushing. The cardboard acts as an insulator and prevents further loss of heat from the nozzle tip. Unfortunately, introducing any foreign material in front of the nozzle may damage its surface, resulting in leakage around the nozzle and perhaps serious damage later.

Clamping Systems

The clamping system of an injection molding machine has one main purpose: to keep the mold closed and under sufficient pressure during the injection of the plastic melt to prevent any plastic from escaping. To understand how this stand how this feat is accomplished, it is necessary to understand the common components of the clamping system.

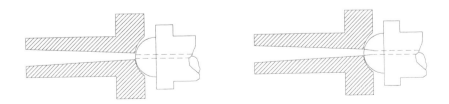

Fig. 6-10 Nozzle tip configurations. (a) Straight (forward) taper. (b) Reverse taper

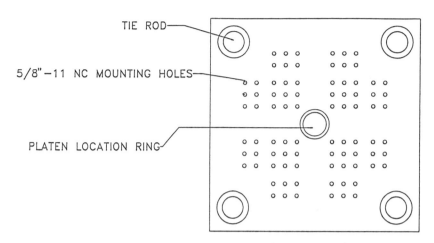

Fig. 6-11 Stationary platen

Platens are the massive cast steel plates that support the clamping system and the mold. They are machined with a specific pattern of holes that allow molds to be easily located and mounted.

The stationary platen (Fig. 6-11) has a 4.000 in. diameter hole, used to align the center of the mold with the center of the platen and, ultimately, the injection unit. The mold has a 3.990 in. diameter locating ring, used to guide the mold into the stationary platen.

The moving platen (Fig. 6-12) has a number of strategically located "through" holes. These are designed to allow knockout bars to pass from the rear of the platen, where they were actuated by a hydraulic cylinder. The knockout bars enter the mold base and bump or push the mold's ejector system forward to eject the plastic part(s).

Both platens have a vast number of tapped holes that are used to hold mounting clamps, which in turn hold the mold halves to the platens. The threaded holes are usually $\frac{5}{8}$ in.–11 UNC for molding machines with clamping force of less than 500 tons. (UNC is one of the three "unified screw thread" designations. UNC stands for coarse, UNF stands for fine, and UNEF stands for extra fine.)

Molding machine manufacturers have been developing a variety of alternative clamping systems to eliminate the need for bolted clamps. These newer systems include rails to slide molds into position, as well as hydraulic locking bars that mount or unmount molds automatically.

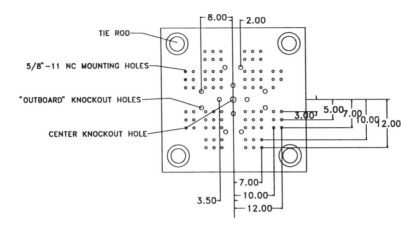

Fig. 6-12 Movable platen

Daylight is the amount of space between platens (Fig. 6-13). Simply stated, maximum daylight is the maximum distance between the platen surfaces when the clamp system is fully opened. Minimum daylight is the minimum distance between the platen surfaces when the movable platen is fully forward without a mold mounted. The maximum and minimum daylight of a molding machine's clamping system is a key differentiating characteristic that tells the processor how large or small a mold can be used in the press, based on the die height of the mold. (The die height is the height of a closed mold, excluding the locating ring. See Chapter 10.) The daylight opening of the molding machine should equal the die height of the mold when the mold is clamped.

Many times a molder would like to install a mold into a press, but the minimum daylight of the molding machine is greater than the daylight of the mold, and such a scenario would prevent the machine from clamping the mold. To accomplish a successful set-up, the molder may add a spacer or bolster plate to augment the mold's die height.

Clamping Pressure. The clamp tonnage required to keep the mold closed during the injection stage is determined by the type of mold, the mold size, and the projected area of the mold (the total surface area of the mold that will be exposed to plastic, as seen through the plane of the parting line of the mold).

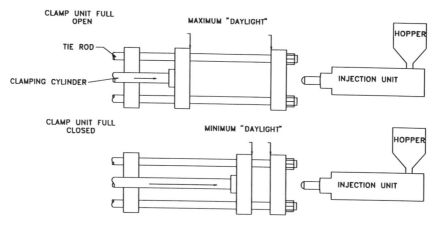

Fig. 6-13 Mold press daylight

Determining the amount of clamp tonnage required for any specific mold is a simple process:

1. *Determine the projected area of the mold.* All areas exposed to plastic material (e.g., cavities, runners, and sprues) should be considered. The depth of a particular part is not a factor.

2. *Determine the tonnage factor.* Usually expressed in tons per square inch, the tonnage factor is a conservative estimate that considers the injection pressures and melt viscosities of different plastics, to facilitate the clamp tonnage calculations. It is developed by plastic material suppliers and by plastics processors through experience. A typical tonnage factor is 3 tons/ in.^{2}.

3. *Multiply the projected area by the tonnage factor.* Example: Assume that in a four-cavity mold, each cavity has a projected area of 5 in.^{2}. The projected area of the total cavity, therefore, is 20 in.^{2}. The total runner system has a projected area of 10 in.^{2}. The total projected area for this mold is 30 in.^{2}. Assuming a tonnage factor of 3 tons/ in.^{2}, the clamp tonnage for this mold would be 30 in.$^2 \times$ 3 tons/ in.^{2} = 90 tons.

Clamping Mechanisms. To effect such a force to clamp a mold, most injection molding machine manufacturers employ one of two basic clamping mechanisms: toggle or hydraulic.

The toggle clamping system (Fig. 6-14) clamps the mold using a mechanical advantage developed through a series of linkages. As the linkages are forced into a straight or closed position by the action of a hydraulic cylinder on a crosshead, the tie bars strain or stretch, and clamping forces are developed. Some of the advantages of toggle systems are their fast motion, low oil flow requirements, and positive clamping action with no bleedoff or pressure loss. On the negative side, they allow the processor little or no control over tonnage variation, and frequent maintenance of linkages and pins is necessary.

The hydraulic clamping system (Fig. 6-15) has made great inroads into the plastics molding machine sector over the past decade. It employs one or more hydraulic cylinders to force the moving platen closed and create the required clamping forces. Hydraulic systems have been around for over 100 years, and the current designs integrate several features to improve efficiency and reliability. Some of the advantages of hydraulic systems are that they have simple designs, allow the processor to vary clamp tonnages, and require little maintenance due to fewer moving parts. The common disadvantages are variations in system pressure, depending on oil viscosity, and inefficiency due to the large volumes of oil moved.

Other Clamping Mechanisms. Although toggle and hydraulic clamping systems dominate the injection molding machine market, several variations of these systems have appeal in certain manufacturing applications:

•   *Hydromechanical clamping systems* are hybrids that use mechanical toggle clamps in the major movement of the platen, but a hydraulic cylinder in the final stages of movement and clamping.

•   *Single-toggle clamping systems,* variations on the standard double toggle, reduce the complexity of the clamping mechanism and therefore the frequency of maintenance.

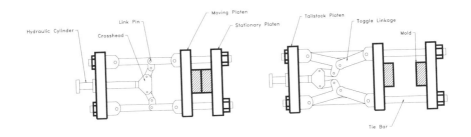

Fig. 6-14 Toggle clamping system

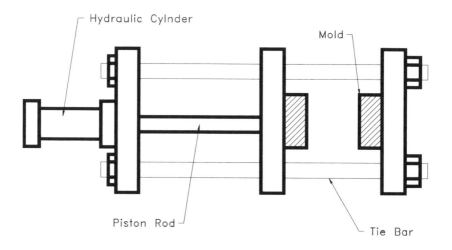

Fig. 6-15 Hydraulic clamping system

- *Electric clamping systems* allow injection molding machines to be used in "clean rooms" for molding precision electronic and medical devices.

Safety

The high clamping tonnages, high injection pressures of hot plastic melt, and number of moving parts demand that injection molding machines have redundant safety features. The mandatory safety features include:

- *Purge guard:* This cover is attached to the injection unit side of the stationary platen. The guard has an electrical interlock that must be engaged before the injection unit can be purged.
- *Front door:* Located on the operator's side, this door prevents operator access to the clamping system during operation. The door has both an electrical interlock and a hydraulic interlock to ensure that the platens cannot move when the door is open.
- *Rear door:* Located on the back side of the molding press, the rear door has an electrical interlock, and sometimes a hydraulic interlock, to prevent the clamp system from operating when the door is open. On smaller molding machines, the front and rear doors may have an additional shield that goes over the top of the press, to prevent long-

armed operators from reaching down into the clamping system during operation.

- *Mechanical safety bar:* This is a fail-safe feature. If all the electrical and hydraulic safety features fail to function properly, the mechanical safety bar will physically retard or stop the forward motion of the movable platen. Depending on its design, this bar may have to be adjusted each time a mold is set up. It is important that set-up personnel and operators alike understand the set-up of the bar.

- *Emergency button:* Located conspicuously on the operator's side of the machine, the emergency stop button is usually designed either to stop all machine functions, or to cause the movable platen to move rearward and then stop all machine functions.

The Injection Molding Process

Temperature Profile

As mentioned above, the injection unit has three main screw zones, each of which is used to progressively melt the plastic pellets. The heater bands on the injection molding machine must be set to a level that provides a similar thermal progression. At the feed zone (rear zone), the temperature controller for the heater bands is set 20 to 30 °F below the melting temperature of the particular plastic being processed. This lower temperature ensures that the screw will be able to convey the plastic pellets in the initial stages of processing. The early ability of the screw to move the pellets is often referred to as "getting a good bite" on the material. If the heater temperature is set too high in the feed zone, the screw will slip, resulting in no conveyance of plastic.

Depending on the size and design of the injection unit, there may be three to five distinct heating zones to control. Assuming that there are four zones, the middle zone may have a temperature setting 10 to 15 °F higher than that of the feed zone, and the temperature of the front zone will be set approximately equal to the desired melt temperature of the plastic being processed. Finally, the nozzle heater is set for a temperature equal to or 10 °F higher than that of the front zone. This is because the plastic within the nozzle receives all its thermal energy from the electrical heater, and the processor wants to keep the plastic melted.

Screw Control

The process variables for the screw depend on the plastic material being processed and the sophistication of the injection molding machine being used. Variables common to most machines include screw speed, pressure, injection rate, shot size, cushion, and screw decompression.

Screw speed ranges from about 50 to 200 rpm, as defined by a tachometer attached to the rear of the screw. Screw rotation is created by an electric or hydraulic motor at the rear of the injection unit. Some motors are connected to the screw with a transmission that allows a higher torque to be created. After the plastic is injected, the screw rotates to convey the plastic forward and provide the necessary shear energy to melt the plastic pellets.

Pressure. There are three main categories of pressure to consider when referring to the injection unit of a molding press: injection pressure, holding pressure, and back pressure.

Injection pressure is also called pack pressure, high pressure, boost pressure, and first-stage pressure. Basically stated, it is the pressure generated by the screw tip on the plastic melt (the pressure used to pack the mold cavity once it has been filled). There is a hydraulic pressure continuum from the screw through the plastic in the mold. Until this pressure is generated, the only pressure on the plastic melt is associated with its resistance as it fills the mold.

One of the most important process control elements of the injection molding machine is to establish at what point mold filling ends and the packing of the cavity begins. This concept is discussed in more detail in the section "Process Control" in this chapter.

Holding pressure is sometimes called hold pressure or second-stage pressure. As its name implies, holding pressure is used to maintain the cavity pressure after the cavity has been packed with plastic. During the time the holding pressure is in effect, the gate cools sufficiently to prevent any plastic from exiting the cavity. Because the plastic material shrinks as it cools, the pressure in the cavity begins to decrease, so the holding pressure can be lower than the injection pressure.

Back pressure is the amount of resistance applied to the rear of the screw as it rotates when conveying plastic. If a constant screw speed is maintained, increasing the back pressure increases the shear force that causes friction and, in turn, melts the plastic. If there is too much back pressure, the turning screw will be unable to return. Too much frictional heat will be generated, and the plastic in the transition and feed zones may melt. If there is too little or no back pressure, the screw will speed back to its set shot size position. There will be a low level of shear and possibly too low a density of the plastic melt in front of the screw. This could cause defects in the plastic parts, such as underfill or sink marks.

Injection rate is the speed of the screw in its ram (plunger) mode. The rate of the forward action of the screw affects the cycle time, shear rate, and cooling of the plastic melt as it is forced into the mold. The electronic control systems available on today's injection molding machines allow the processor not only to precisely control the injection rate, but also to program it. Microprocessors, which give the molding machines a certain intelligence, and a sensor, which determines the exact ram position, allow the injection rate to vary within the injection cycle.

Older injection molding machines had injection rate selection options of fast, medium, or slow. Subsequent experimentation and computer analysis of plastics mold filling showed that consistently high-quality parts can be molded if the injection rate can be varied during the time the screw is moving forward. Now processors can change the injection rate as many as 16 times during a 2 sec injection time. The machine can inject quickly when the sprue and runner of the mold are being filled, then slow down as the plastic passes through the narrow gate into the cavity. To use such a tool effectively, the processor must thoroughly understand the mold filling process.

Shot size is the distance the screw is allowed to move rearward during rotation. Therefore, it is related to the amount of plastic that is deposited in front of the screw. The precision of shot size control is a function of the sensor on the machine that determines the position of the screw. A linear potentiometer is widely used to accomplish this task. Such a sensor is accurate within 0.001 in.

Cushion is the plastic melt that remains in front of the screw after the plastic melt has been injected. Most thermoplastic injection molding machines leave a cushion to ensure that the screw does not "bottom out" against the head of the machine, and as a safety margin in case there are shot size adjustments to be made. On some molding machines, the amount of cushion depends on the shot size. For example, if a larger cushion is required, the shot size is increased, and the material that remains in front of the screw after injection is the cushion. Most modern machines use a separate cushion control circuit.

Screw decompression, sometimes called screw suckback or pullback, is a machine control feature that allows the plastics processor to pull the screw rearward a predetermined distance after the shot size distance has been achieved. When the screw is pulled back, it is not rotating; therefore, no additional plastic melt is deposited in front of it. As a result, the material already in front of the screw has extra space, which relieves the pressure on the material. This keeps the material from drooling out of the nozzle prematurely and increases process controllability.

Process Control

Process control aims to reduce variation within a specific process. It is dependent upon the machine's ability to have its variables accurately set and precisely controlled. It wasn't until the microprocessor was developed and integrated into injection molding machines that processors had the ability to maintain such repeatability (Fig. 6-16).

Theories of how to control the injection process vary from manufacturer to manufacturer and are interpreted differently by different plastics processors. The one area of agreement is that the events that take place within the injection unit, during the injection phase of the molding process, are the focus of control.

Fill-to-Pack Transition. As discussed above, when the ram travels forward, the plastic melt begins to fill the mold. As the mold becomes full, the duty of the injection ram changes from simply conveying the plastic melt to packing the plastic melt into the mold cavity by applying high pressure. Once the fill-to-pack transition point has been determined empirically, the next objective is to control the consistency of the transition from cycle to cycle. The technique

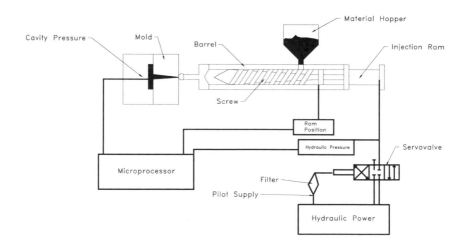

Fig. 6-16 Microprocessor-controlled injection molding machine

for this is based either on time, ram position, hydraulic pressure, or cavity pressure.

Time. The assumption underlying this control technique is that all aspects of the ram motion are consistent and that if a timer is set, the packing portion of the process will always take place at the same point. This is unrealistic, because the injection speed varies with changes in melt viscosity and hydraulic oil pressure. Nevertheless, this technique was used by several processors prior to the addition of the microprocessor.

Ram Position. The assumption underlying this control technique is that if the transition point is always allowed to occur at the same point as the ram moves forward, the plastic part being molded will be consistent in form and dimensions. Ram position is one of the most common process control techniques employed by machine manufacturers, not because it is the most popular among processors, but because it is one of the least expensive features to add to a machine. It is also very simple to use.

Hydraulic Pressure. A more complex technique specifies the hydraulic pressure for the ram, with little or no consideration of ram position. The assumption is that if there is control over the hydraulic pressure of the plastic in the injection unit, there will be equivalent control of the hydraulic pressure of the plastic within the mold cavity. Hydraulic control of the fill-to-pack transition is often confusing to the inexperienced operator, however, and this technique has not been as popular as ram position control.

Cavity Pressure. In this technique, the cavity pressure within the mold is monitored, and when it reaches a predetermined level, the fill-to-pack transition occurs. This form of process control is a true closed-loop system: the events that occur within the mold are sensed and reported to the microprocessor, which controls the injection ram. If the mold cavity sensors determine that the pressure, speed, or point of transfer from fill to pack is inadequate, adjustments are made to the screw variables.

There is no doubt that it's most realistic to control the injection process and the fill-to-pack transition based on the actual events that are occurring within the mold cavity. After all, the plastic part is produced within the cavity, so why not always control the machine by monitoring cavity pressure? The answer is that cavity pressure control requires the processor to modify molds to accept a cavity pressure transducer, an extra cost and effort that many processors would rather avoid. Additionally, the molding machine must have a cavity pressure control system, which is usually also an added cost and tends to be avoided. In production quantities, the added expense of cavity pressure control is readily justified.

Process Documentation. To document the injection process, the most widely used technique is to identify and mark as many parameters as possible on a set-up sheet, such as the one depicted in Fig. 6-17. This technique is not consistent, however, because the process used to make quality plastic parts will vary as a function of such factors as:

- Resin (color, regrind, and batch)
- Hydraulic oil temperature
- Moisture and environmental conditions
- Operator involvement with the process
- Equipment condition and maintenance

Therefore the trend within the plastics industry is to *program* set-up parameters, using the computer-based control system available on modern molding machines. The set-up sheet is replaced with a set-up diskette, card, or file. This

Product Part Number	Product Name		Mold Number	
Plastic Material Part No.	Plastic Material		Part Weight	
Clamp Size Min. Max.	Shot Size			
Pressures	**Temperatures**		**Timers**	
Injection	Injection Unit		Injection Unit	
Profile A	Rear		Ram Forward	
Profile B	Transition			
Profile C	Front			
	Nozzle			
Holding				
Profile A	**Mold Related**		**Clamp System**	
Profile B	Heating Medium W / E/ O		Mold Closed	
Profile C	Water Pattern Specification			
	Stationary			
Back	Moving		**Speeds**	
	Mold Dieheight		Injection Rate	
Positions	K.O. Bar Size		Screw Speed	
Injection Unit	Eye Bolt Size		Mold Slow Close	
Shot Size	Core Pull Y/N		K.O. Forward	
Decompression	Multi-Eject Y/N No.		K.O. Return	
Cushion	Mold Weight			
K.O. Stroke	Mold Orientation H/V			
Clamp System			**General**	
	Transitions		Packing Specification	
Mold Open-Slow	Fill to Pack			
Mold Open-Fast			**Mold Maintenance**	
Mold Open-Slow			Cleaning	
Mold Close-Slow			Lubrication	
Mold Close-Fast				
Mold Protection				

Fig. 6-17 Set-up documentation sheet

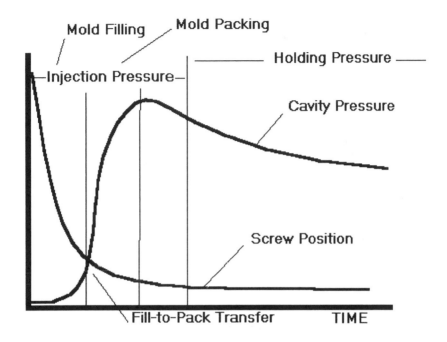

Fig. 6-18 Process "fingerprint"

form of programmable set-up ensures that the molding process will be set up the same way consistently, and that changes to the process can be validated and controlled. Most machines can log the specific times when changes are made, monitor how long it takes a set-up to occur, and sound alarms when an out-of-process condition exists.

Process "Fingerprinting." Prior to microprocessor-controlled machines and process instrumentation, molders had to trust that if the machine was set up correctly, the process events would occur consistently. Still, they knew this was an ideal that simply did not exist. Today, injection process variables such as mold cavity pressure, injection pressures, and fill-to-pack transitions can be monitored in real time and be documented for further reference. The magnitude of pressure, including mold cavity pressure, along with the times at which mold filling and mold packing occur, can be recorded in a process "fingerprint" (Fig. 6-18). If a process is producing quality parts, the fingerprint can be used as a reference for future process control. Because the objective of process control is to reduce

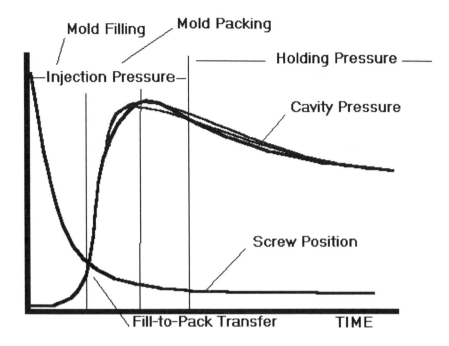

Fig. 6-19 Multiple process "fingerprints" showing that the process is in control

variation, the molder can define how much variation is allowed, if any, and measure and adjust the process as required.

Multiple fingerprints can be seen in Fig. 6-19 and 6-20. Figure 6-19 represents a process that is in control: several consecutive tracings are overlapped, reflecting narrow variation in the process. Figure 6-20, on the other hand, illustrates a process that is not in control: the tracings do not overlap as closely, reflecting broad variation between cycles. Such a process would not produce consistently high-quality parts.

The search for an optimum process control system can be very challenging for small and midsize plastics processors. Because a significant number of injection molding machines in use have no process control features, the marketplace abounds with several varieties of add-on process controllers. The confusion is no less for plastics processors who purchase new equipment with integrated process control systems.

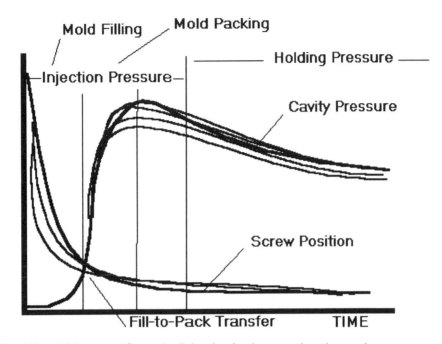

Fig. 6-20 Multiple process "fingerprints" showing that the process is not in control

One important suggestion for processors: Do capability studies on a variety of molding machines or control systems before you buy one. Have the new machine or processor used on your production floor. This way the system can be used by your technicians and you can seek their invaluable opinions. Additionally, attach the system to a pressure profile printer and fingerprint the process variations. It is not unreasonable to evaluate a system over a three-month period and to evaluate four systems over a year. Such evaluations are key to ensuring that the system will perform as expected, and they are prudent because of the large capital investment involved.

Thermoset Injection Molding

Thermosetting plastics are often not considered candidate materials for injection molding, but this is a misconception. Injection molding of thermoset-

Fig. 6-21 Thermoset injection molding screw. Courtesy of Spirex

ting plastics such as phenolics, polyesters, and epoxies can result in excellent strength, temperature resistance, and stability, properties not attainable with most thermoplastic materials. Most thermosetting plastics behave like a thermoplastic material for a short period of time, which provides a sufficient window of opportunity for them to be processed by the injection process.

The injection molding machine used for thermoplastic injection molding must be modified to allow thermosets to be processed. In most cases, injection presses specifically designed for thermosetting plastics are preferred. Most of the differences between thermoset injection presses and thermoplastic injection presses are found within the screw and barrel of the injection unit.

The screw used for a thermoset injection machine is visibly different from a typical thermoplastic screw (Fig. 6-21). Most noticeable is the absence of a check valve at the discharge end. Because the check valve increases the shear and friction, so that extra heat is generated, thermosetting plastics would tend to cure (solidify) if it were present. Without a check valve, the screw cannot be allowed to rotate during injection, so a brake on the rear of the screw is used to prevent rotation during injection. The compression ratio of the thermoset screw is directed to a 1:1 augur, so little work is done by the screw to mix the plastic and generate heat.

The barrel of a thermoset injection molding machine has its temperature controlled by water controllers similar to those used to control thermoplastic mold temperatures. A typical barrel for thermoset machines has three or four zones that are set to temperatures in the range of 170 to 230 °F. This temperature range keeps the thermosetting plastic cool enough to prevent barrel cure.

Other Differences. Thermoset injection molding machines do not use a cushion of plastic in front of the screw. All the thermoset material is injected into the mold, to reduce the risk that plastic will cure in the nozzle tip. In addition, to ensure a sufficient cure, the molds used for thermosetting materials must be heated to between 275 and 375 °F, using electricity or oil. (The clamp systems used on thermoset injection presses are identical to those used on thermoplastic presses.)

Fig. 6-22 Thermoset injection molding machine. Courtesy of Bucher, Inc.

At first glance, thermoset injection machines resemble thermoplastic molding machines (Fig. 6-22); however, they are specifically designed to meet the challenges of processing thermosetting plastics. These machines and their associated processes can be readily learned by technicians who are familiar with processing thermoplastic materials. Some of the complex parts moldable on today's sophisticated thermoset molding machines are highlighted in Fig. 6-23.

Future Trends

Injection molding will continue to be the largest and fastest-growing component of the plastics processing industry. The emphasis will be on efficiency improvements, such as set-up time reduction, better process control, and reduction in energy use. Further work will be in the development of small desktop molding machines that will spend their manufacturing lives producing one part at the point of use. These machines will allow true just-in-time manufacturing by totally eliminating dedicated processing areas. Efforts are also underway to produce more complex parts with the injection process, using multiple materials and in-mold decorating, with the goal of producing a complete plastic system, not just single plastic parts.

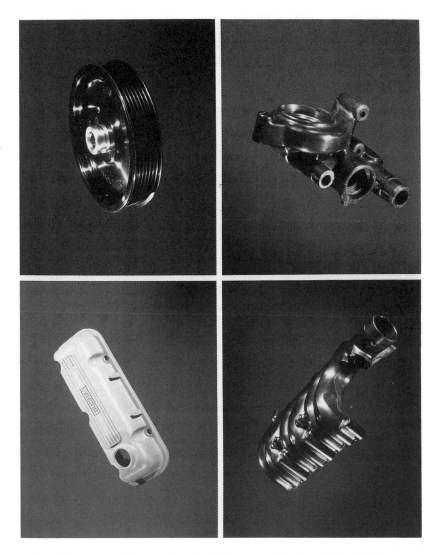

Fig. 6-23 Examples of parts produced with thermoset injection molding. (a) Valve cover. (b) Phenolic pulley with aluminum insert. (c) Intake manifold. (d) Water pump housing. Courtesy of Bucher, Inc.

7

Composites Processing

Composite Plastics

Composite plastics are defined as materials that contain at least two phases, one a polymer matrix (binder) and the other a reinforcing fiber, particulate, or flake. The reinforcements, usually glass fibers, are used to significantly improve the strength of the plastic. The term *fiberglass* is a general term for a thermosetting polyester plastic matrix reinforced with glass fibers, strands, or cloth. Thermoplastic materials that are reinforced may also be considered composites. Thermoplastic composites are readily molded using conventional plastics processes such as injection molding, extrusion, and blow molding.

The composite system is defined by:

- The type of reinforcement
- The amount of reinforcement
- The matrix (polymer) used

An example is 30% glass-filled epoxy. Composite materials allow the part designer and processor to customize a plastic compound by adjusting the type and volume of reinforcements added to a specific plastic matrix. Such composites usually exhibit synergistic behavior: the strength of the composite is

greater than the sum of the individual strengths of the matrix and the reinforcements. Other advantages of composite plastics are that they:

- Provide better load distribution than conventional plastics, which
  results in a superior strength-to-weight ratio
- Can be used in large parts
- Result in excellent product durability
- Offer superior chemical resistance

Although the development of composite materials and fabrication methods has made great strides over the past 20 years, there are still very few
high-volume processes available to manufacturers. Additionally, because
composite product fabrication is limited, there are few design rules available.
(In contrast, designers of products fabricated with conventional processes,
such as injection molding, blow molding, extrusion, and thermoforming, can
use a series of guidelines based on other designers' experience with the
materials and processes involved.) Other disadvantages of composite processing include:

- High product cost
- High cost and low availability of equipment
- Low production rates

The processes discussed in this section include compression molding,
structural reaction injection molding, resin transfer molding, sheet forming,
contact molding, pultrusion, and filament winding.

Compression Molding

Compression molding is one of the oldest plastics processes. Originally
used to produce plastic parts made from phenolic, melamine-formaldehyde,
or epoxy, compression molding is seeing a comeback with the development of
new thermoplastic and thermosetting materials.

Compression molding machines consist of two platens, one stationary and
the other movable (Fig. 7-1, 7-2). The principle of operation is simple: The
molding compound is either manually or automatically loaded into the mold
cavity, which is located on the lower platen to facilitate the loading process.
Once the molding compound is loaded, the mold is closed. The force associated with the mold closing, along with the heat of the mold, causes the

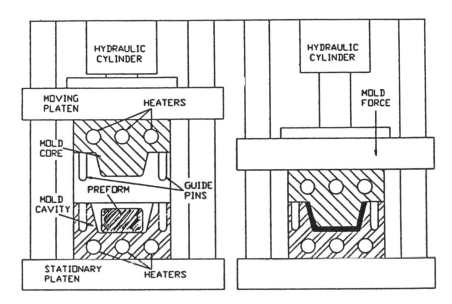

Fig. 7-1 Compression molding

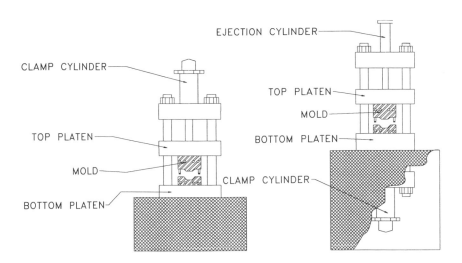

Fig. 7-2 Compression mold presses. (a) Upper platen movement. (b) Lower platen movement

molding compound to flow for a short period of time prior to curing. (The potential force of these machines ranges from 10 to 10,000 tons.) The mold may remain closed until the plastic is cured; then the mold opens and the part is ejected.

The condition of the material prior to its being loading into the cavity varies from a dry powder to a preheated preform or pill. The advantage of compression molding is that the plastic material flows for a short period of time and does not need to be forced through narrow runners, gates, or cavities. A short flow length reduces the shear stress, allowing the creation of a product with high dimensional stability.

Structural Reaction Injection Molding

Structural reaction injection molding (SRIM) is in its technological infancy. It is simultaneously undergoing dramatic materials and process improvements and rapid industrial growth. SRIM is a very attractive composite manufacturing process for producing large, complex structural parts economically. The main drawback is that because it is relatively new, some aspects of the technology are poorly understood, and some crucial equipment remains to be fully developed.

The basic concepts of the SRIM process are shown in Fig. 7-3 and 7-4. A preformed reinforcement is placed in a closed mold, and a reactive resin mixture is impingement mixed under high pressure in a specially designed mix head. Upon mixing, the reacting liquids flow at low pressure through a runner system and fill the mold cavity, impregnating the reinforcement material in the process. Once the mold cavity has filled, the resin quickly completes its reaction. A completed component can often be removed from the mold in as little as one minute.

In many ways SRIM is the natural evolution of two more established plastic molding processes, reaction injection molding (RIM) and resin transfer molding (RTM, discussed later in this chapter). SRIM is similar to RTM in employing preforms that are preplaced in the cavity of a compression mold to obtain optimum composite mechanical properties. It is like RIM in its intensive resin mixing procedures and its reliance on fast resin reaction rates. The term *structural* was added to the term *reaction injection molding* to indicate the more highly reinforced nature of the typical composite components manufactured by SRIM.

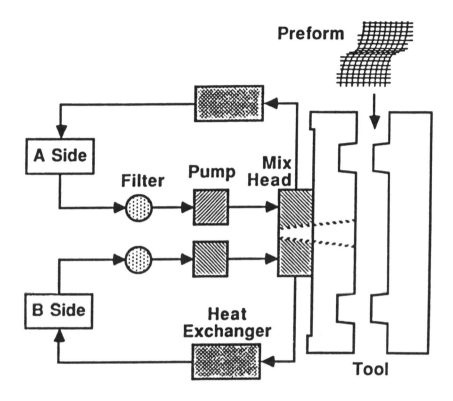

Fig. 7-3 Structural reaction injection molding (SRIM). Courtesy of Shell Development Center

SRIM Preforming

The key to all molding processes that use liquid composites (SRIM, RTM, and cold pour molding) is the preform, a preshaped, three-dimensional precursor of the part to be molded. It does not contain the resin matrix; it can consist of fibrous reinforcements, core materials, metallic inserts, or plastic inserts. The reinforcements, cores, or inserts can be anything available that meets the economic, structural, and durability requirements of the parts. This tremendous manufacturing freedom allows for a large number of alternative preform constructions.

Preform Materials. Most commercial SRIM applications have been in general industry or in the automotive industry, and the reinforcement material

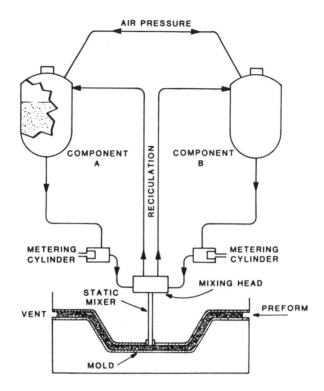

Fig. 7-4 Another view of structural reaction injection molding (SRIM). Courtesy of Ford Motor Co.

most commonly used has been fiberglass, due to its low cost. Fiberglass has been used in the form of woven cloth, continuous strand mat, or chopped glass.

Space-shaping cores can be used in the SRIM process to fabricate thick, three-dimensional parts with low densities. Fiberglass reinforcements and inserts can be placed around these cores, resulting in SRIM parts that are structurally strong and stiff, can be molded in one piece, and are very light-weight. Specific grades of urethane-based foams, those that have densities of 6 to 8 lb/ft^3 and dimensional stability at SRIM molding temperatures, are commonly used as molded core materials. For applications requiring a core that can withstand high compressive loads, end-grain balsa has proven to be suitable.

Metallic inserts have been used in SRIM parts as local stiffeners, highly stressed attachment points, or weldable studs. For the majority of these inserts, the metallic material of choice has been steel.

Molded plastic inserts have been used in some SRIM applications for very specific functions. Velcro strips can be molded on the surface of SRIM parts to mechanically hold other parts in place. Extruded plastic tubing can be used in SRIM parts as internal conduits for electrical wires.

Preforming Processes. While it is easy to visualize a SRIM preform and make a limited number of them by hand, it has proven challenging to make high volumes of identical preforms economically. Preforms for RTM and cold pour molding have been made since the 1950s, either by directed fiber preforming (DFP) or by a manual cut-and-sew process. For low production volumes or for lightly stressed applications, these preforming techniques are suitable for SRIM. More mechanized preforming processes are currently under development for high production volumes.

Directed fiber preforming (DFP) is a partially mechanical method for making chopped-glass, three-dimensional preforms. A stream of chopped fiberglass and binder is sprayed onto a perforated, three-dimensional metal shaping form through which a high volume of air is being drawn. The resulting fiberglass preform replicates the shape of the perforated screen.

Two distinctive types of DFP machines have been developed: plenum chamber and open spray. They are similar in that a resinous binder and chopped-glass fibers are collected onto a screen by air movement. The binder is then cured through the application of heat, and the preform is removed from the screen.

A *cut-and-sew preform* is assembled much as a piece of clothing is assembled on a dummy. Patterns representing the individual sectional layers of the part are used as templates to cut pieces of the preform from a roll of the reinforcing material. Woven fiberglass cloth, unidirectional mat, and/ or continuous strand mat are commonly used in this process. Either these cut pieces of fabric are placed onto a forming shape, or they are placed directly into the cavity of the SRIM mold and mechanically stapled to the core. Additional reinforcing fabric layers can then be added as necessary. Metallic or plastic inserts can also be added to the preform. Eventually this preform is placed into the cavity of the SRIM mold, and resin is injected into it.

Needless to say, this is a very labor-intensive process, and the consistency from preform to preform is usually poor. However, for very low manufacturing volumes this process can be cost-effective. It is almost always the procedure used to make preforms for prototype SRIM parts.

SRIM Resins

The commercialization of a large family of low-viscosity, fast-reacting, economical resins was responsible for the industrial debut of SRIM. The activities of resin suppliers such as Amoco, Arco, Ashland, BASF, Dow, Goodrich, Hercules, ICI, Miles, and OCF have speeded up the adoption of SRIM by many molders. Most SRIM resins have several characteristics in common:

- Their liquid reactants have room-temperature viscosity below 200 cps.
- Their viscosity-cure curves are sigmoidal in shape, and the typical mold fill time is 10 to 90 sec (Fig. 7-5).

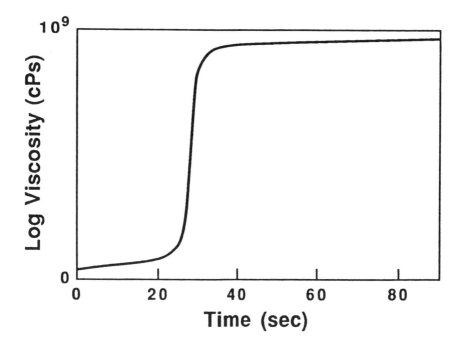

Fig. 7-5 Viscosity-cure behavior of a typical SRIM resin. Courtesy of Shell Development Center

• Their demold time is from 60 to 180 sec, varying with catalyst concentration.

The low viscosity of SRIM resins and their relatively long fill times are crucial in allowing them to penetrate and flow through their reinforcing preforms. If their viscosity is too low (<30 cps), the turbulent flow in the mold cavity can entrap air and create voids in the composite. However, some turbulent flow through the reinforcement mats is desirable, because it serves as an after-mixing process leading to a more homogeneous cure. Excessive viscosity can retard flow through the preform, leading to resin "racetracking." This causes large voids, dry spots, and/or distortions in the preforms, which when cured become the major SRIM molding defects.

SRIM Equipment

The equipment and processing techniques used by SRIM have largely evolved from those developed for RIM, RTM, and reinforced reaction injection molding (RRIM), except that those related to preforming tend to be unique to SRIM. As illustrated in Fig. 7-6, SRIM does occupy a molding pressure/cycle

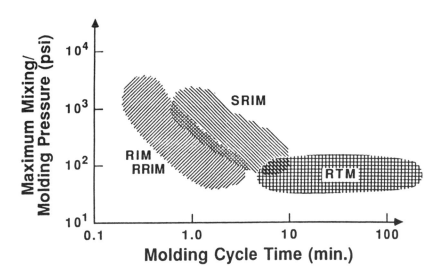

Fig. 7-6 Molding pressure/cycle time windows for reaction injection molding/reinforced reaction injection molding (RIM/RRIM), structural reaction injection molding (SRIM), and resin transfer molding (RTM). Courtesy of Shell Development Center

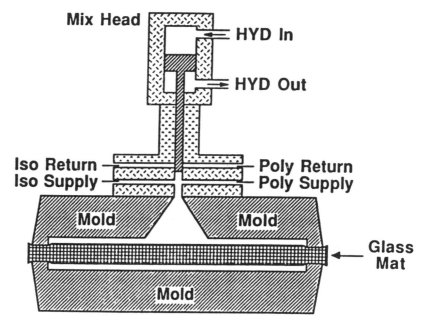

Fig. 7-7 Cross section of SRIM equipment. Courtesy of Shell Development Center

time window that is slightly different from those occupied by RIM/ RRIM or RTM.

The increased use of SRIM can be attributed, in part, to the fact that most RIM/ RRIM equipment can be used in SRIM. However, high-volume SRIM applications usually involve the development of more specialized molding equipment. Sources for complete RIM/ RRIM/ SRIM systems include Admiral Equipment, BASF/ Elastogram Machinery, Battenfeld of America, Cannon USA, Cincinnati Milacron, Krauss-Maffei, and Perros (EFY Corporation).

A growing trend in RIM/ RRIM/ SRIM equipment is modularization. Using this approach, the equipment builder constructs a unique SRIM unit composed of individual modules that meet the customer's requirements. These modules can be replaced with enhanced units as the needs of the molder change or as technology advances.

Figure 7-7 is a cross section of a typical SRIM molding equipment configuration. SRIM equipment is composed of two equivalent sets of mechanical components for handling the active A and B sides of RIM resins. The individual resin components are stored in tanks under a positive pressure, usually provided by a dried, inert gas. The resin is then drawn through the filters to remove any contaminants that might clog the fine orifice of a downstream mix head.

Impingement pressure for the resin components in the mix head is provided by axial piston pumps or lance cylinders. The mix head, mounted to the outside of one cavity of the compression mold, provides injection/mixing action, which initiates the polymerization reaction and fills the cavity of the mold with the proper amount of reacting resin. When the mixing head is not injecting resin into the mold in its high pressure mode, it recalculates the A and B reactive resin components back into their holding tanks. Usually the RIM injection units also have a low-pressure mode in which recalculating A and B components bypass the mix head.

A heat exchanger is used by each side of the SRIM equipment to regulate the temperature of the resin components. Usually, optimum property and processing results can be achieved by maintaining resin component temperatures at or slightly above the ambient room temperature (23 °C). The heat exchanger generally functions as a cooling device: The resin travels through fine orifices in the SRIM mix head during its recalculating path, and this mix head is mounted on a heated mold cavity, so the resin components can quickly become hotter than desired.

SRIM Molds. Once the resin has been mixed, the overriding processing concerns are to fill out the mold and wet the reinforcing preform without distorting it, all in a matter of seconds. The design of SRIM molds, their composition, and the techniques used to ensure complete mold filling are critical components in completing SRIM mold making.

Like SRIM parts, SRIM molds have become more complex in the last five years. While molds with moving slides and ejector pins are used extensively in thermoplastic injection molding, the very low viscosity of hot, reacting SRIM resins makes it difficult to seal these moving slides and ejectors in SRIM molds. In addition, the glass reinforcement used in most SRIM preforms is abrasive and can cause peening of the sealing surfaces of a mold. Eventually, fiberglass containing flash (which stoutly resists trimming) forms in these areas.

In the past, most high-volume SRIM applications used steel molds. With the trend in many industries toward more frequent part design changes and subsequently shorter production runs per mold, a slow shift toward nonsteel mold materials is underway. The low pressures encountered in SRIM molding are encouraging this movement. (Some molders do suggest that when SRIM production volume exceeds 10,000 parts, molding should usually be done in steel tools, depending on the nature of the part being molded.) The chief alternatives to tool steels for use in SRIM molds are:

- Castable steel-like alloys
- Kirksite (an alloy of aluminum, copper, zinc, and traces of other metals)
- Chemically bonded ceramics
- Cast and machined aluminum
- Epoxy
- Sprayed metal

Table 7-1 compares these materials for use in SRIM molds. Kirksite, chemically bonded ceramics, and epoxy generally are less expensive than the other tool materials mentioned above. However, poor thermal conductivity of these materials usually bars their use in high-production SRIM runs, where a rapid resin heat-up rate is desirable to minimize cycle time. In prototype part applications, this is usually not a significant factor, and these materials are widely employed.

SRIM mold filling occurs rapidly. Generally it is measured in seconds, even for parts with high glass loadings. To assist in directing the flow of resin into the preform of large SRIM parts, runners are often machined into the surface of a mold. The resin will find the path of least resistance as it flows through a mold cavity packed with a preform, so its flow route is often very circuitous. The placement of anisotropic preforms, preforms of various densities, and/ or

Table 7-1 Materials for use in SRIM molds

Tool material	Equivalent injection molding tool costs, %	SRIM use
Machined mold steel	80-100	Production
Machined aluminum	60-80	Production
Castable "steel-like" alloys	50-70	Production
Kirksite tooling	45-65	Limited production, prototype
Cast aluminum	35-55	Limited production, prototype
Chemically bonded ceramic	30-50	Very limited production, prototype
Sprayed metal	25-45	Very limited production, prototype
Nickel shell epoxy	25-35	Very limited production, prototype
Epoxy	15-30	Prototype

impervious inserts in a complex, three-dimensional mold cavity can lead to resin racetracking.

The design of the gating and runner configuration (if any) is usually kept proprietary by the molder. However, it appears that most SRIM parts are center gated, with vents located along the periphery of the part. This configuration allows the displaced air in the mold cavity to be expelled uniformly. Deviations from this approach are unlikely except when the molder encounters problems with racetracking or poor mold filling. These mold modifications are usually based on the experience of the molder and on trial-and-error experimentation.

Some molders use the RTM technique of sealing the edges of a tool and pulling a vacuum through the vent ports as an aid in filling out SRIM molds. Others have found that this procedure only encourages racetracking. Recently, detailed observations of SRIM mold evacuation have demonstrated that this procedure decreases the size of entrapped bubbles during molding, but ultimately increases the maximum pressure during fill. Similar studies, which have measured the permeability of continuous strand mats, unidirectional mats, and bidirectional mats in multiple layers, have provided insights into synergistic flow effects that can be used to optimize mold fill-out.

Understanding these phenomena is more important as SRIM parts become larger and more complex in shape and as the fiberglass volume fraction approaches 30%. Eventually a point is reached where injection through a single port cannot fill the mold before the resin gels. The solution is to use multiple injection ports, shot either simultaneously or sequentially. The optimum placement of the injection ports and the optimum timing of their injection can be determined through computerized mold filling simulation.

SRIM Applications

The first commercially produced SRIM part was the cover of the spare tire well in several automobiles produced by General Motors. Since then, SRIM automotive structural parts have included foamed door panels, sunshades, instrument panel inserts, and rear window decks. Nonautomotive SRIM applications include satellite dishes and seat shells for the furniture market.

The ability of SRIM to fabricate large, lightweight composite parts, consisting of all types of precisely located inserts and judiciously selected reinforcements, is an advantage that other competitive manufacturing processes find difficult to match. In addition, large SRIM parts can often be molded in 2 to 3 min, using clamping pressures as low as 100 psi. Thus, the capital requirements of SRIM are relatively low, allowing economical manufacture of parts when annual production volumes are below 10,000 units. These advantages, coupled with the concurrent development of a large family of commercially available resins, have led to forecasts of a high annual growth rate in SRIM.

If SRIM is to achieve its potential, the development of economical preforming and manufacturing processes is essential. While SRIM resin suppliers have developed a wide array of potential composite matrices, little is known about their compatibility with existing fiberglass reinforcements, their long-term stability in potentially degrading service environments (e.g., temperature extremes, chemicals, high humidities, and fatigue loading), or ways to improve their surface appearance.

Nevertheless, there has been significant progress in SRIM since the first commercial part was produced in 1985. Further development will not require revolutionary discoveries; rather, application of existing technologies from other industries will lead to a fully unified and economical SRIM manufacturing process.

Resin Transfer Molding

Resin transfer molding (RTM) is similar to SRIM in that it is well-suited for the manufacture of large, complex, high-performance polymer composite structures. As with any manufacturing technique, a thorough understanding of the advantages and limitations of the process is required before it can be considered for the design and manufacture of a component.

In its common form, RTM is a closed-mold, low-pressure process in which dry, preshaped reinforcement material is placed into a closed mold and a polymer solution or resin is injected at a low pressure, filling the mold and thoroughly impregnating the reinforcement to form a polymeric composite part. The reinforcement and resin may take many forms, and the low pressure, combined with the preoriented reinforcement package, provides a large range of component geometries, sizes, and performance options.

Figure 7-8 illustrates the RTM process. A preform of reinforcement fibers, in this case a stampable glass mat, enters the process as a roll of flat sheet material and is stamped with a shaping die. The preform is preassembled with foam cores in three-dimensional box sections, if desired, and then the finished preform is placed in a mold and the mold is closed. A two-component resin system is mixed in a static mixer and metered into the mold through a runner system. Air inside the closed mold cavity is displaced by the advancing resin front and escapes through vents located at the high points of the mold, the last areas to fill. When the mold has filled, the vents are closed, the resin inlet is

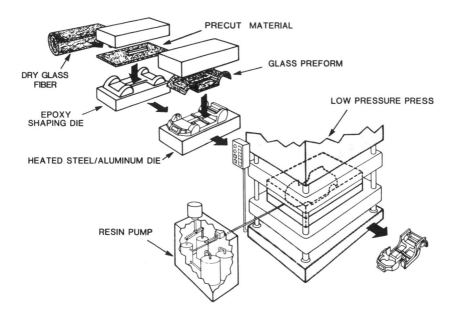

PRECUT MATERIAL

GLASS PREFORM

DRY GLASS
FIBER

LOW PRESSURE PRESS

EPOXY
SHAPING DIE

HEATED STEEL/ALUMINUM DIE

RESIN PUMP

Fig. 7-8 Resin transfer molding (RTM). Courtesy of Ford Motor Co.

closed, and the resin within the mold cures. Upon complete cure of the resin, the finished component is removed from the mold.

The most elementary form of RTM is depicted in Fig. 7-9. A simple, two-dimensional preform of glass mat is formed into a concave mold cavity, and the upper half of the mold is closed on the reinforcement. The outer perimeter of the mold is left open to allow the displaced air to escape as the resin fills the mold from the center to the edges. As the resin reaches the edges of the part, it simply drains out into a catch basin. When injection is complete, the component is allowed to cure, then is removed from the mold and trimmed. The resulting physical properties approach those of even the most sophisticated RTM systems.

A more sophisticated RTM system, sometimes referred to as high-speed resin transfer molding (HSRTM), is shown in Fig. 7-10. This process uses a complicated preform, one or more of several high-speed preforming processes, and usually foam cores, molded-in inserts, and high-performance attachment points. Tooling is designed for high volume and rapid cycle times,

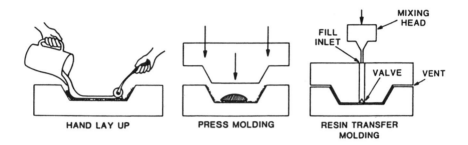

Fig. 7-9 Elementary form of RTM. Courtesy of Ford Motor Co.

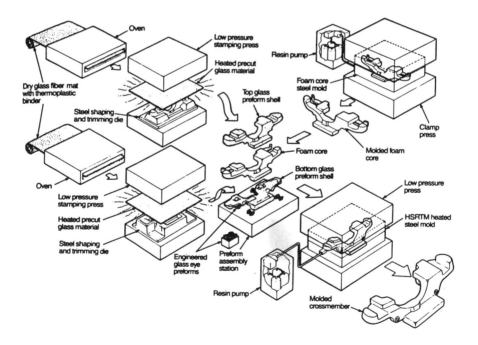

Fig. 7-10 High-speed resin transfer molding (HSRTM). Courtesy of Ford Motor Co.

and it is normally of steel or nickel shell construction for maximum durability. Resin is introduced using automated resin handling equipment, and the resin inlet and air outlet (vent) are accomplished with automated self-cleaning nozzles.

HSRTM uses the same resin system employed in the slower RTM process, but the chemical reactivity of the system can be increased to yield cycle times of 1 to 6 min with heated tooling. The use of highly automated equipment makes HSRTM very similar to SRIM.

RTM Prototyping

RTM is an excellent process choice for making prototype components. It allows representative prototypes to be molded at low cost, unlike processes such as compression molding and injection molding, which require tools and equipment approaching production level in order to accurately simulate the physical properties achievable in the production-level component. In some cases, RTM can be used to make prototype components for other processes, and the RTM component will typically have properties that exceed the production-level product.

When prototyping with RTM, less-reactive resins are generally used, allowing long fill times and easier control of the vents. Tooling is usually low-cost epoxy, but it could be made with an impervious material that would contain the resin. Prototype preforms are made by cut-and-sew methods, and any foam cores used are machined to shape. Sizes can range from small components to very large, complex, three-dimensional structures. RTM provides two finished surfaces and controlled thickness, and it requires no auxiliary vacuum or autoclave equipment. Other processes used for prototyping, such as hand lay-up and wet molding, give only a single finished surface, and dimensions in the thickness direction are controlled.

RTM Applications

Today, the majority of RTM structures fall into two categories: low-volume structures where RTM is used to manufacture large, relatively complex parts; and small, simple geometries that are manufactured in low to medium volumes. The challenge for the composites industry will be to extend the potentially useful RTM technology into higher-volume markets, areas now dominated by metals. In the transportation sector, this extension will be driven by the need for lower-weight structures that increase fuel efficiency and by the need for increased styling flexibility at acceptable investment levels.

Outer Panels for Automobiles. An exploded view of the current concept of composite outer panels for automobiles is shown in Fig. 7-11. The clear preference emerging from recent developments is for compression molded materials, such as sheet molding compounds (SMC), to be used for horizontal panels, and for RIM materials or thermoplastic injection molded materials to be used for vertical panels. Horizontal panels require significant stiffness, reasonably low coefficients of thermal expansion, and a Class-A surface finish. Vertical panels need the high toughness of thermoplastics to provide "friendly" features such as dent resistance. In addition, the quality of thermoplastic-based materials is comparable to that of steel, because the processing techniques for thermoplastics produce outstanding surface finishes.

The approach envisioned in Fig. 7-12 is identical to the SMC approach, except that the two-piece SMC assemblies would be consolidated into single RTM structural elements. RTM could also consolidate the entire body structure into a single piece. The economics of manufacture might dictate that the structure be divided into additional pieces, but there would be significantly

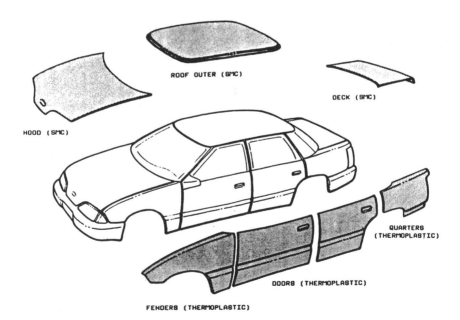

Fig. 7-11 Current approach to outer panels for automobiles. Courtesy of Ford Motor Co.

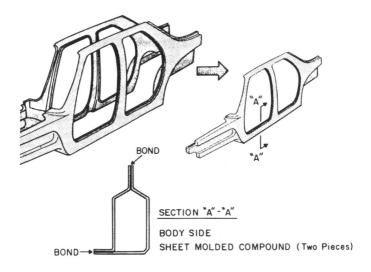

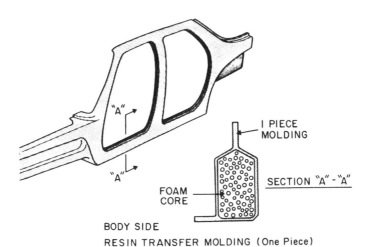

Fig. 7-12 Envisioned approach to outer panels for automobiles. Courtesy of Ford Motor Co.

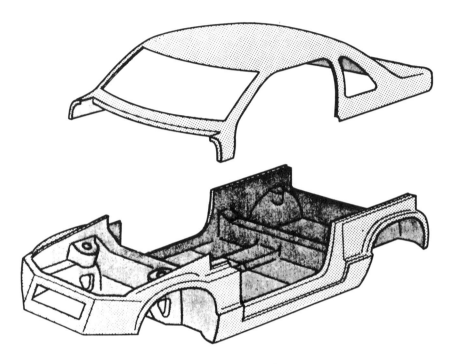

Fig. 7-13 Envisioned approach to full-body automobile structure, using RTM. Courtesy of Ford Motor Co.

fewer pieces than in the SMC approach. Other anticipated benefits of RTM over the SMC approach are that:

- Tooling costs and assembly costs would be lower.
- RTM pressures would allow the molding of larger components, and they would lower the energy requirements of the process.
- Because of the orientation of the reinforcing fibers, structural integrity would increase and there would be higher overall stiffness at lower weights.

The ultimate use of composite materials in a vehicle structure would be for RTM to produce a single, large, integrated body structure. Figure 7-13 envisions this structure as a single piece, but once again, the economics of manufacturing might dictate that it be subdivided into several structural modules. In this concept, all structural functions are performed by the composite, including management of crash energy. No Class-A finish would be required, because additional surface panels could cover the exterior of the chassis.

Recycling of RTM Components

Recycling of composite structures has not been a major issue, due to the low number of structures being produced. However, an effective resource recovery strategy will ultimately be needed for large RTM components, because they normally have mineral filler content and some level of high-dollar-value reinforcement. This may provide some incentive for development of recycling processes. Absence of filler in the system will provide scrap that has a higher energy content and less residual noncombustible materials, possibly making RTM material more attractive for incineration.

Thermoplastic Composite Sheet Forming

The term *advanced thermoplastic composite* refers to highly loaded reinforced thermoplastics, generally over 50% fiber by volume. Because of their use in applications requiring high specific strength and modulus, they are made of layers of aligned continuous or long discontinuous fibers with specified orientations. The reinforcing fibers may be carbon, Kelvar,® or glass (Table 7-2).

Some of the most important potential advantages of using advanced thermoplastic composites instead of thermosets are the ability to automate the process and the lower cost of converting intermediate material forms into end-use parts. The superior processability of the thermoplastic component eliminates the long and expensive process steps associated with thermosets.

One of the most attractive processing techniques for converting thermoplastic composites to parts is sheet stamping. Cold sheet stamping is a low-cost, fast process that is widely used in the sheet-metal forming industry. It has been applied in various forms to unreinforced plastics, usually in high-temperature processes. Examples of these are vacuum forming, solid-phase high-pressure forming, and hot stamping, such as is used in the processing of random glass fibers in polypropylene.

In a typical thermoforming process, the sheet stock, or preform, is heated in an external oven or in a forming system (Fig. 7-14). At the forming temperature, the sheet is transferred into dies and shaped to conform to the die geometry. After forming, the sheet is cooled under pressure to below the resin glass transition temperature (T_g) before it is removed from the forming system.

Depending on the preform used, the part generally has to be trimmed to yield the final net shape, so materials waste considerations are important in the design of the preform and the deformation path during forming.

Although the forming techniques used for advanced thermoplastics are similar to those used for sheet metal and unreinforced thermoplastics, there are some important differences between these materials:

Thermoset versus thermoplastic processing

Thermosets	Thermoplastics
Low viscosity	High viscosity
10-100 poise	1000 poise
Low temperatures	High temperatures
120-200 °C	300-400 °C
Low pressures	High pressures
Vacuum-150 psi	100-2500 psi
Long curing times	Fast heating and cooling rates(a)
Autoclave capital investment	Autoclave not required(a)
Refrigeration	No refrigeration
"Wet" process	"Dry" process

(a) Biggest processing advantages

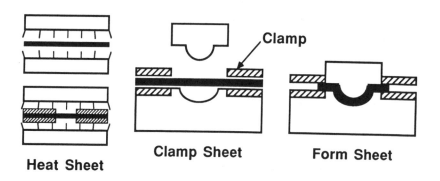

Heat Sheet   **Clamp Sheet**   **Form Sheet**

Fig. 7-14 Thermoforming using matched dies. Courtesy of DuPont Engineering Research and Development Division

Table 7-2 Commonly referenced advanced thermoplastic composite forms

Trade name	Supplier	Resin	Resin type	Fibers	Fiber form	Sheet structure	$T_g/T_m/T_{proc}$, °C(a)	Comments
APC-2	ICI	PEEK	Semicrystalline	Carbon glass	Continuous	Uni-tape	144/340/380	
HTX	ICI	PAK	Semicrystalline	Carbon glass	Continuous	Uni-tape	205/386/420	
LDF™	Du Pont	PEKK	Semicrystalline	Carbon glass, Kelvar®	Discontinuous	Multi-ply sheets, woven ribbon tow fabric	156/338/370	
Ryton®	Phillips Petroleum	PPS	Semicrystalline	Carbon glass	Continuous	Uni-tape, woven fabric and ribbon tows	85/285/343	
…	Quadrax	Various	Semicrystalline	Carbon glass, Kelvar®	Continuous	Woven ribbon tows	…	
…	BASF	PEEK, PPS, PEI	…	Carbon glass	Continuous	Powder impregnated, commingled fibers, uni- and woven fabrics	…	Unconsolidated, tacky
Filmix®	Heltra/ Courtaulds	PEEK	Semicrystalline	Carbon glass	Discontinuous	Commingled/ co-spun twisted yarn uni-tape and woven fabric	144/340/380	Unconsolidated
CYPAC®	American Cyanamid	PEI	Amorphous	Carbon glass	Continuous	Uni-tape, woven fabric	217/—/343	
RAYDELC®	Amoco	PAE	Amorphous	Carbon	Continuous	Uni-tape, woven fabric	260/—/350	
RAYDEL-X	Amoco	PAE	Amorphous	Carbon	Continuous, woven fabric	Uni-tape	220/—/350	
PAS-2	Phillips Petroleum	PAS	Amorphous	Carbon glass	Continuous	Uni-tape	215/—/330	
HTA	ICI	PES	Amorphous	Carbon glass	Continuous	Uni-tape	260/—/330	

(a) T_g, glass transition temperature; T_m, melting temperature; T_{proc}, processing temperature

These differences affect how the materials are actually applied:

- Forming of advanced thermoplastic composites is done at much higher temperatures and pressures than that of thermosets and unreinforced thermoplastics.
- Due to the high viscosity, flow processes take longer in thermoplastic composites, and the heating and cooling times are greater because of the higher temperatures.
- Maintenance and control of fiber placement and orientation during forming, required in thermoplastic composites to retain high properties, makes the deformation path more critical than in isotropic unreinforced thermoplastics and sheet metal.
- Although forming techniques are similar for thermosets and thermoplastics, more sophisticated solutions are needed for successful processing of thermoplastic composites.

Table 7-3 compares the forming characteristics of unreinforced thermoplastics, sheet metal, and advanced thermoplastic composites.

Table 7-3 Typical forming characteristics

	Material		
Characteristic	**Unreinforced thermoplastics(a)**	**Sheet metal**	**Advanced thermoplastic composites(a)**
Sheet heating	IR, convection	None	IR, convection
Sheet characteristic at process temperature	Pliable, drawable	Drawable	Drawable in discontinuous fiber systems only
Sheet structure	Isotropic	Isotropic	Anisotropic
Composition	Homogeneous	Homogeneous	Heterogeneous (two-component)
Process temperature	Medium ($T_g + 50$ to 75 °C)	Low (room temperature)	High (280-390 °C)
Cycle time	Short (minutes)	Very short (fractions of minutes)	Long (tens of minutes)
Deformation loads	Low	Medium to high	Medium to high
Consolidation pressure	Low (vacuum)	None	High (100 to 1000+ psi)
Mold temperature	Low (below T_g)	Room	High (sheet temperature)
Compensation for thickness variations in matched dies	Yes	Yes	Not easily

(a) IR, infrared radiation; T_g, glass transition temperature

Sheet Forming Materials

One of the most important decisions in sheet forming is the choice of a sheet structure. At this time, the main concern is availability, because manufacturers make most sheet materials available only in developmental quantities. The general classifications are woven fabrics, unidirectional fiber tapes (also called uni-ply), and random-fiber sheets. When the first two are plied together in multi-angle, multi-ply configurations, the resulting sheets are referred to as multi-ply laminates. The resin properties determine processing temperature, and the amount of fiber structure distinguishes the sheets from a processing standpoint.

When thermoplastic composites were introduced, the tendency was to manually lay them up like thermoset prepregs (reinforcement material that has been impregnated with a resin matrix). Initially this was difficult because the thermoplastic prepregs were stiff and hard to handle, but the new generation of thermoplastics allows for such easy lay-ups that they rival thermoset composites.

Sheet Forming Techniques

In general, the method of sheet forming determines the quality of the part and the cost of producing it. The choice depends on what level of capital investment is desired and how the deformation path of each method will affect the fiber distribution of the resulting part. Table 7-4 compares the major processing methods currently used in forming.

Table 7-4 Sheet forming methods and equipment

Method	Advantages	Disadvantages
Vacuum forming	Low pressure	Low force, impractical
Matched metal die press	Close tolerance, high forming load and pressure	Thickness mismatch, friction at die interface, nonuniform deformation/pressure, long/heating/cooling times, high fabrication cost
Hydroforming	Only one solid die, undercuts high pressure	Low temperature, no peripheral equipment
Diaphragm forming	Good fiber placement control	Long cycle time, temperature limitations
Rubber pad forming	Only one solid die, high pressures	Lack of complex details, temperature limitations

Matched-die press forming is probably the most widely used forming system. Forming presses are readily available and vary from small hand-operated presses to fairly sophisticated computer-controlled hydraulic systems. For simple forming operations, standard heated-platen presses used for flat panel molding are adequate. However, in operations where control of deformation rate and pressure history are important, high-quality stamping presses are used.

The dies used in matched-die press forming are generally made of metal, which can be internally heated and/or cooled. When metals are used, the dies are generally designed to a fixed gap (thickness) of close tolerance. High pressures can easily be applied to the workpiece. A disadvantage of this forming method is that when there is a thickness mismatch between the formed piece and the premachined cavity, nonuniform pressure is produced on the part, resulting in nonuniform consolidation.

When heating or cooling is desired, the dies usually have such a high heat content that heat transfer times are long. In addition, matched-die fabrication costs are high because the two close-tolerance die halves must match. Substituting an elastomeric material for one of the die halves usually reduces the cost and permits application of a more uniform consolidation pressure than is possible in an all-metal die set.

Diaphragm forming is arranged so that the workpiece is held between two disposable, plastically deformable diaphragms. The diaphragms are then clamped, heated with the workpiece to the processing temperature, and deformed over a tool half, using a combination of air pressure and movement of the tool in contact with the workpiece (Fig. 7-15). The workpiece is usually under a vacuum throughout the process.

The workpiece is under compression normal to the surface, and the actual deformation rate is slow (controlled by the creeping of the diaphragm), so control and lateral fiber motions are good, at least for parts with simple geometries. Because the diaphragm material deforms in a finite temperature range, only materials that can withstand these temperatures can be used, and this limits the geometric complexity of the part. Typical diaphragm forming temperatures are 300 to 400 °C for materials such as DuPont's Kapton polyimide film. In addition, cycle times for current forming systems are long (90 to 120 min) due to heating and cooling of the massive pressure chambers used to provide the pneumatic forming pressures. However, cycle time has been reduced to as short as 60 min, using specifically designed low thermal inertia tooling and chamber preheating.

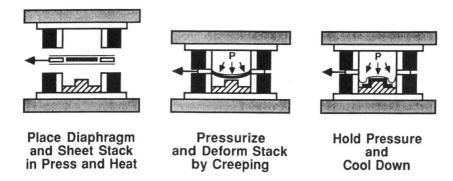

Place Diaphragm   Pressurize   Hold Pressure
and Sheet Stack   and Deform Stack   and
in Press and Heat   by Creeping   Cool Down

Fig. 7-15 Diaphragm forming. Courtesy of DuPont Engineering Research and Development Division

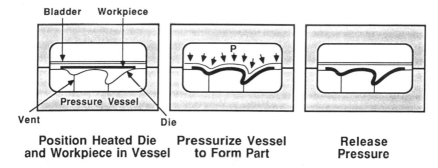

Position Heated Die   Pressurize Vessel   Release
and Workpiece in Vessel   to Form Part   Pressure

Fig. 7-16 Hydroforming. Courtesy of DuPont Engineering Research and Development Division

Hydroforming is similar to diaphragm forming in that a fluid medium, usually hydraulic fluid behind a rubber diaphragm, is used to deform the workpiece against a male or female tool half (Fig. 7-16). The main differences are that the rubber diaphragms in hydroforming are a permanent part of the forming system and are usually much larger than the workpiece. Only one diaphragm is used, but because the rubbers used in these systems are suitable only for lower temperatures, disposable rubber sheets are placed over the workpiece to prevent diaphragm rupture due to high temperatures or sharp points on tools.

Hydroforming systems have been used extensively in the sheet-metal forming industry. They are usually massive systems capable of up to 10,000 psi. The high pressures allow undercuts to be formed, but the disposable rubber sheets, which can be as thick as ½ to 1 in., limit how small the cavities can be and still obtain tool definition. The tooling can be heated using the tool bed that is moved into the forming area. Because the whole system is pressurized, it is impractical to use peripheral devices such as clamps or cooling and heating lines connected to the tool itself.

Autoclave/Vacuum Forming. Autoclaves, which have traditionally been used for thermoset composite curing, are now being used increasingly often for thermoplastics. They provide a natural extension for thermoset molders who are interested in developing thermoplastic know-how by using equipment that is already installed. The principles of autoclave processing of thermoplastic composites are not too different from those of hydroforming or diaphragm forming. The cavity between the workpiece and the tool is generally evacuated or exposed to atmospheric pressures to create the pressure differential necessary to deform and consolidate the workpiece (Fig. 7-17). If the pressure in the autoclave is atmospheric and the evacuated cavity is under a vacuum, the process is essentially just vacuum forming. For thermoplastic composites, vacuum forming is impractical because higher pressures are generally needed for deformation and consolidation within a reasonable time.

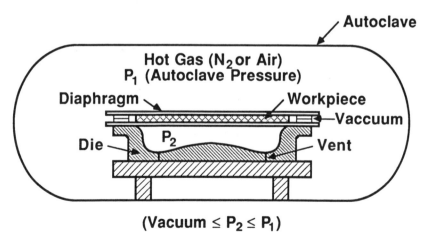

Fig. 7-17 Autoclave/ vacuum forming. Courtesy of DuPont Engineering Research and Development Division

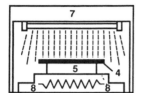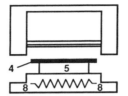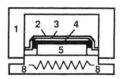

Fig. 7-18 Rubber pad forming at elevated temperature. (a) Heating outside the press. (b) Table moved into the press. (c) Pressing. 1, Rubber pad retainer. 2, Rubber pad. 3, Wear pad. 4, Blank. 5, Form block. 6, Die table. 7, Infrared blank heater. 8, Table heaters. Courtesy of DuPont Engineering Research and Development Division

Rubber pad press forming involves replacement of one tool half by a thick pad of rubber that conforms to the solid tool half under pressure in a forming press (Fig. 7-18). The rubber pad can be profiled to the tool geometry, but it remains permanently attached to the press platen and is generally much larger than the tool. This contrasts with rubber-matched die tooling, where the rubber is matched to the metal tool half.

Rubber pad press forming provides high forming pressures that are not uniform over the workpiece. The pressures are determined by the local extent of deformation, and they approach uniformity for parts of very shallow draw. This limits the complexity of parts that can be formed with good detail. To obtain a good level of draw, the bulk of the rubber pad is made of a compliant foam material, such as polyurethane, with a cover pad of silicone rubber to enable high-temperature forming.

Sheet Forming Tools

More often than not, the design of the forming tool and the choice of tooling materials do not get much attention during process specification. Yet forming dies ultimately determine the geometric accuracy of formed parts, and in some cases where the product lot is small, they account for a significant portion of the part fabrication cost.

Due to the high temperatures and pressures required to process advanced thermoplastic composites, the choice of materials is limited. In general, dies are made from steel, aluminum, metallic alloys such as Kirksite (castable) or Invar, or castable or chemical-bondable ceramics. All of these materials have limitations when applied to thermoplastic composite sheet forming (Table 7-5).

Table 7-5 Tooling materials and limitations

Tool material	Type	Fabrication method	Cost range	Max temp, °F	Dimensional, stability	Resistance to impact	CTE in/in °F (10^{-6})
Aluminum alloy	Metal	Machine or cast	Medium	600	Fair	Fair	
Steel	Metal	Machine	Medium high	1000	Good	Good	
Invar	Carbon, nickel, steel alloy		Medium high				1.5-6.0
Kirksite	Zinc alloy	Cast	Medium	715	Good		
Comtek™ (OxyChem)	Ceramic	Chemically bonded		800			4.0
CeraCom (C/SiO_2) (Com Tool Tech)	Carbon/ ceramic	Heat curable @350 °F (pre-preg)		2000			0.225
Silicone rubber	Elastomer	Cast		600	Poor	Fair	
Polyimide	Carbon composite	Hand lay-up		750			
Epoxy	Carbon	Cast or hand lay-up		250-350			2.5

The Sheet Forming Process

The key processing steps in sheet forming are: heating the sheet to be formed, placing it inside dies, shaping it by forming, and cooling it under pressure to consolidate it.

Sheet Preparation/Heating. Typical sheet preform heating methods include infrared radiation (IR), conduction (e.g., heating between two heated platens), and convection (in an air or inert gas circulation oven). For IR, heating time is usually short (of the order of 1 to 2 min), but for thick sheets, temperature gradients develop through the thickness. Depending on the sheet thickness, conduction heating can be configured to provide heating times similar to those of IR, but heated sheets tend to stick to the contact surface, making handling difficult. Convection heating usually takes the most time (5 to 10 min), and the use of inert gas is preferable, just as in IR heating, to prevent oxidation of the polymer at high temperatures.

Transfer to Dies. Sheets to be formed are usually heated to the melt temperature of the resin. At this temperature, the sheets are very flexible, deconsolidated (if not heated under pressure), and easily distorted. This makes handling difficult, especially for large, thin sheets, if they are heated outside the forming system and have to be transferred. In this case, some form of framing is required to hold the sheet for transfer into the forming die. Transfer times are of the order of a few seconds to prevent significant cooling.

Sheet Forming/Fiber Placement. The deformation step in a forming process is just a way to redistribute the fiber and resin in a composite over a three-dimensional curved surface. Success in making an acceptable part depends on placing the fibers and resins at specific locations in order to attain the required properties. The ability to control fiber placement is one of the most important criteria in the selection of the forming system. In general, for continuous fibers, lateral fiber displacements must occur to accommodate the increasing or decreasing surface area associated with going from a flat surface (sheet) to a curved, three-dimensional surface.

Sheet Forming Applications

The market for large thermoplastic composite parts is increasing, with the military and the aerospace industry leading the process and product trends. Typical products include helicopter wing sections and aircraft access doors. The automotive sector is working with thermoplastic composite systems for the fabrication of large structure panels.

The key limitations of sheet forming technology are:

- Lack of commercial quantities, and high cost, of sheet preforms
- High up-front development costs, because sheet forming behavior cannot be accurately predicted
- High tooling costs
- Lack of quality control techniques to measure fiber placement and ensure consistent quality
- Long cycle times for processing due to high temperatures
- Joining techniques
- Lack of extensive data on formed part performance and viscoelasticity

Contact Molding

When composites are fabricated with contact molding, the process is performed under conditions appropriate for thermoset, room-temperature-cure resins. Usually a single mold surface is involved, although these molds can become very large or complex in shape and may use breakaway segments to allow for undercuts. The majority of the parts are completed using manual compaction, and very few exceed 50 psi.

Contact Molding Processes

The three major groupings of contact molding processes are hand lay-up, spray-up, and vacuum bag molding. They differ in the means of applying resin, distributing the reinforcement, and ensuring densification and air release.

Hand lay-up involves manual placement of cut-to-size reinforcement onto the mold surface (Fig. 7-19). The resin may be dispensed by catalyst injection using a spray gun, or precatalyzed resin may be dispensed with a spray gun or a napped paint roller. Distribution of the resin and final compaction are usually completed manually. The operations may be mechanized singly or in part by special-purpose saturators, rollout devices, or robotic systems, depending on the part shape and the productivity required.

Spray-up uses fiberglass in roving form, drawn through an air-motor-driven chopper (Fig. 7-20). The chopper is usually mounted above a spray gun that dispenses the resin. The same type of spray gun used in hand lay-up may be used. The combined system is usually manually operated and is capable of dispensing laminate at upward of 25 lb/ min. Thus, labor control and crew balancing are important in operations such as tub/ shower manufacturing, where this is the predominant process.

The spray lay-up process may be used on complex mold shapes, but because the fiber is applied in random fashion, the upward limit for fiber content is approximately 38%, and parts are intermediate in strength. However, spray lay-up can be used with more oriented fibers via the hand lay-up process. Considerable quantities of reinforcement are consumed in this market segment. Recent industry estimates indicate that at least 40% of all E-glass reinforcement used in the contact molding portion of the composites industry is applied by spray lay-up. Thus there has been a continuing effort to improve the performance of the materials and equipment.

Vacuum bag molding shares some of the attributes of the hand lay-up process (Fig. 7-21). However, because it frequently uses heated molds, ovens, or autoclaves for matrix curing, additional material such as prepregs may be used, and very high mechanical properties may be realized. Vacuum bag molding is often the best process when relatively simple, high-structural-performance and small tooling investments are required. A wide range of materials in prepreg form are available, although they generally do not include

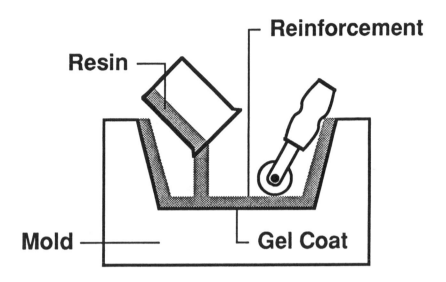

Fig. 7-19 Hand lay-up. Courtesy of EMZ Associates

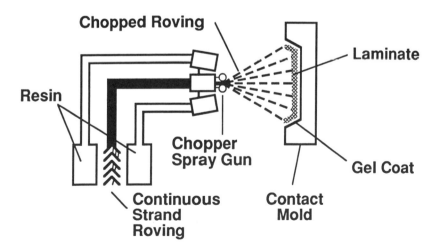

Fig. 7-20 Spray-up. Courtesy of EMZ Associates

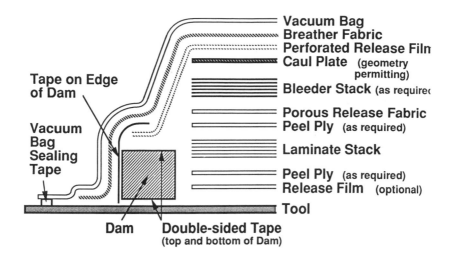

Fig. 7-21 Vacuum bag molding. Courtesy of EMZ Associates

chopped strand fibers. Oriented and woven near-net-shape materials are widely available in all commercial fibers.

Design constraints arise because of the nature of the vacuum compaction process and the material that is needed to effect it. In general, parts cannot be as thick as those molded in a single mold cycle by hand lay-up, but there are techniques to overcome this in specific applications, such as ballistic armor parts. Alternative techniques for laminate build-up by successive, partial curing episodes are possible if epoxy matrix materials are used. The mold preparation time can be very lengthy on short production runs because of the nature of the vacuum process and the sealing and release systems required. However, this can be substantially reduced with prefabricated vacuum blankets and other accessories for more extended production.

The production rate in vacuum bag molding depends on the complexity of materials placement, as in hand lay-up and spray-up, and on the added effort required for vacuum materials placement. Partial curing of secondary material, if necessary, will take several hours, followed by materials placement and another cycle on the same mold. Thus, complex parts of very high structural capability may take several days to complete. One part per mold or shift is probably a conservative minimum in the absence of a specific part configuration.

Table 7-6 General design considerations for hand lay-up and spray-up

Characteristic		Hand lay-up	Spray-up
Minimum inside radius		1/4 in.	1/4 in.
Molded-in holes		Large	Large
Trimmed-in mold		Yes	Yes
Undercuts (split mold)		Yes	Yes
Minimum draft recommended		2°	2°
Minimum practical thickness		0.030 in.	0.060 in.
Maximum practical thickness		Virtually unlimited	Virtually unlimited
Maximum thickness variation		±.020 in.	±.020 in.
Maximum thickness build-up		As desired	As desired
Corrugated sections		Yes	Yes
Metal inserts		Yes	Yes
Surfacing mat		Yes	Yes
Limiting size factor		Mold size	Mold size
Metal edge stiffeners		Yes	Yes
Bosses		Yes	Yes
Fins		Difficult	Difficult
Hat sections		Yes	Yes

(continued)

Table 7-6 General design considerations for hand lay-up and spray-up

Characteristic		Hand lay-up	Spray-up
Molded-in labels		Yes	Yes
Raised numbers/letters		Yes	Yes
Translucency		Yes	Yes
Strength orientation		Random or directional	Random
Typical fiberglass reinforcements (loading percent by weight)		25-65%	20-50%

Contact Molding Design

General design considerations for contact molding are discussed in detail in texts on reinforced plastic and in information supplied by the major vendors of reinforcement materials. An overview for hand lay-up and spray-up appears in Table 7-6. In general, contact molding provides the widest latitude in part size and shape, the limitations being the molder's ingenuity in placing the reinforcement and removing the part from the mold (or vice versa).

One finished surface is representative of the quality of the mold surface and is usually colored with a polyester gel coat. Surfaces can weather very well and can provide a complete range of colors. Surface texture is available if the mold surface is so configured. Undercuts, ribs, molded-in inserts, molded-in holes, and section changes (by selective reinforcement or core materials placement) are also available. Mechanical properties can be relatively high, as fiber contents of 55 to 65% can be achieved, depending on fiber type and orientation. Although polyester resin is the predominant material, vinyl ester, epoxy, and special formulations of phenolic are also available. New materials include interpenetrating network systems, urethanes formulated for spray-up, and hybrids.

Contact Molding Applications

Contact molding has few limitations of size or configuration, and this flexibility has led to increased penetration into large-scale projects. As traditional materials wear out and prove to be energy-inefficient, fiber-reinforced plastic (FRP) systems have an opportunity for significant additional growth.

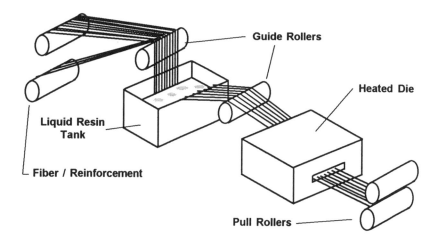

Fig. 7-22 Pultrusion

The marine market has been the major market for contact molding processes for decades. It remains the largest in terms of long-fiber structural applications and the use of resins and gel coats. In recent years there has been an accelerating trend for the domestic fishing industry to convert to FRP boats, as wooden boats deteriorate and metallic boats become more costly to maintain, and this trend has been mirrored somewhat overseas. The United States is currently building a class of FRP mine sweepers, although the ship was designed in Italy. The continued mine sweeper program may establish confidence in processes and materials, as well as provide an opportunity for the evaluation of newer systems.

Land Transportation. The movement of goods by truck and rail will require replacement fleets of more energy-efficient carriers. Contact molding has significantly penetrated the truck market, and E-glass composites such as shells, skins, and core are heavily represented in trailer applications.

Infrastructure Replacement and Expansion. The United States, along with most other industrialized countries, is outgrowing and wearing out its people-supporting systems. Composites are being considered for applications once produced with concrete and steel, including drainage pipes, bridge fascia, and even building repairs.

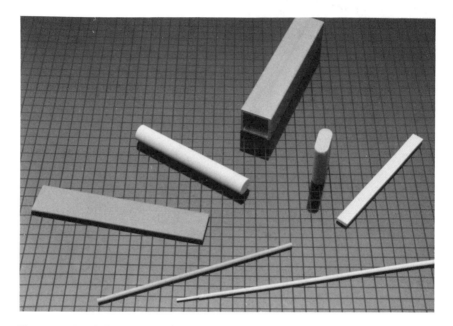

Fig. 7-23 Pultruded poles and shafts. Courtesy of Coastal Engineering Products, Inc.

Pultrusion

Pultrusion has many similarities to extrusion, as the names suggest (Fig. 7-22). It begins with strands of reinforcement material, usually glass or carbon fibers, that have been wetted in a resin tank. The resin used is most often polyester or an epoxy. The next step is to pull the resin-soaked strands through a heated shaping die, which may be a rod, tube, I-beam, or other geometric shape. The resulting profile has a high strength-to-weight ratio and is very durable, especially in a chemical environment.

Current applications for pultruded products include poles and shafts (Fig. 7-23) and structural beams for electrical and chemical environments (Fig. 7-24), such as ladders that can be used near electrical wires. The future of pultrusion may be in space applications, where long structural components for space

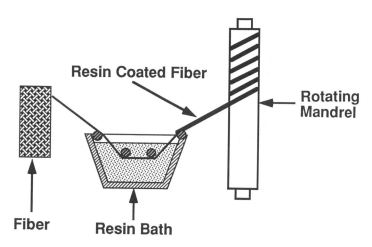

Fig. 7-24 Pultruded structural beams. Courtesy of Coastal Engineering Products, Inc.

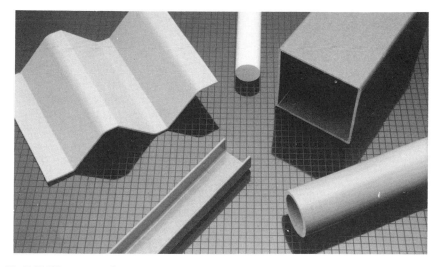

Fig. 7-25 Filament winding

stations could be manufactured in space, as required, eliminating the need to transport long beams from Earth.

Filament Winding

Filament winding is primarily used to manufacture large structural containers or tanks. The process involves directing several spools of reinforcing material, such as glass or carbon strands, into a resin bath of either polyester or epoxy. The wetted strands are wound over a turning mandrel in different patterns to provide different strengths (Fig. 7-25). After the resin has cured, the part is removed from the mandrel and is machined or assembled as required.

Some key applications for filament wound composites are gasoline storage tanks, septic tanks, large-diameter drainage pipes, chemical storage systems, and sporting equipment, such as golf club shafts and bike frames.

8

Thermoforming, Resin Casting, and Recycling

Thermoforming

Thermoforming, also referred to as vacuum forming, begins with plastic sheet as the starting material. The plastic sheet is placed into a clamp frame to hold it securely on all edges (Fig. 8-1). The sheet material may be placed into the clamp frame manually, robotically (for high-volume processing), or continuously (if the sheet material is produced by an in-line extruder). Thermal energy, usually in the form of convection and radiant heat from electrical heating elements, is applied for a sufficient amount of time to soften (not melt) the plastic sheet.

Once the sheet is sufficiently softened, a mold is brought into contact with the sheet, and a vacuum is applied that draws the softened sheet toward the mold to mirror the configuration of the mold (Fig. 8-2). After the sheet cools and the mold is removed, the sheet retains the shape of the mold. Figure 8-3 shows a basic thermoformer.

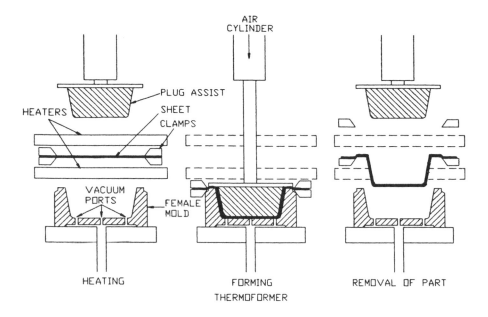

Fig. 8-1 Thermoforming (vacuum forming)

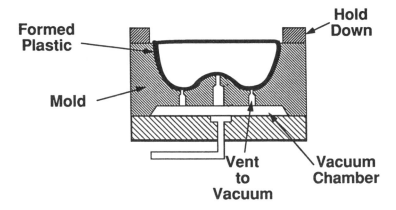

Fig. 8-2 Detail of mold in thermoforming

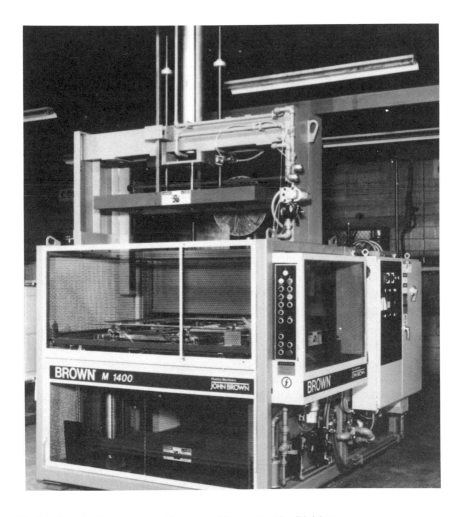

Fig. 8-3 A typical thermoformer. Courtesy of Brown Machine Division

Historically, thermoforming has been considered a one-sided process. The softened sheet mirrors either a male mold, so that the inside is the critical surface and the outside is the noncritical surface, or a female mold, so that the outside is the critical surface and the inside is the noncritical surface. This approach to thermoforming was satisfactory for decades, while the process was used primarily for simple packaging parts. Today, thermoforming has advanced to include molded products that have two critical sides and require close dimensional accuracy so they can be used in key automotive and building and construction applications.

Thermoforming Tooling

The thermoforming process offers some unique tooling advantages over other conventional plastics processes, primarily because thermoform molds are relatively simple in design and construction. Many designers insist on a product design review that includes one or more thermoformed prototypes, to allow a reasonable assessment of product form, fit, and function. Using the thermoforming process, the part designer and manufacturer can make simple prototype molds quickly, using inexpensive materials. In addition, many manufacturers produce their initial production runs on low-cost thermoforming molds until the product matures. Later, the product design is embellished and requires injection molding.

The tooling materials for prototype thermoform molds include wood, plaster, gypsum, and epoxy. The tooling materials acceptable for low-volume production, under 500 parts, include epoxy, beryllium copper, and aluminum. Most high-volume production molds for the thermoforming process are produced using cast and/or machined aluminum.

Tooling designs vary, but some of the standard thermoforming mold designs are male molds, female molds, plug-assisted female molds, and matched molds (Fig. 8-4). The thermoform mold also includes a number of small holes (0.032 to 0.063 in.). These are strategically machined on and around the mold details to provide the vacuum or negative pressure required to form the softened plastic sheet against the mold surface. Also required is a vacuum distribution box to ensure that the vacuum is distributed to all the vacuum holes in the mold.

Thermoforming Expense

Thermoforming is one of the least expensive plastics processes. The equipment required is relatively simple in design and operation, and the tooling is

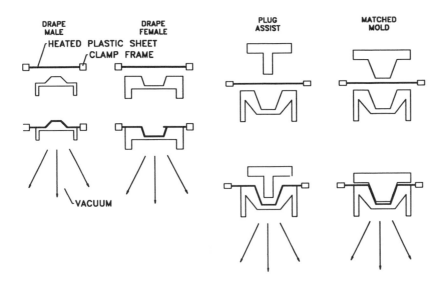

Fig. 8-4 Standard thermoforming mold designs

simple and low-cost, compared to the tooling for other plastics processes, because of the low pressures involved and the low abrasion of the plastic.

The greatest expense in thermoforming is for the materials, from the extruded plastic sheet that serves as raw material to the wasted material generated because the plastic sheet must be held in a clamping frame. This material can be regranulated and reintroduced into the sheet extrusion process from which it was made. Labor is also more costly than for other plastics processes because operator involvement is greater; the operator must often handle raw material or trim the formed sheet.

Thermoforming has two distinct markets, each with their own cost structures. The high-volume packaging market uses thermoformers that are integrated directly with sheet extruders, and the product is automatically cut from the formed sheet stock (Fig. 8-5). This form of thermoforming is lower in cost. The lower-volume nonpackaging market, which uses the thermoforming process because of its ability to produce large plastic parts, is significantly higher in labor and materials costs.

Fig. 8-5 Automatic in-line thermoforming for the packaging market

Typical Thermoformed Parts

The majority of thermoformed products are produced for the packaging market, but there are broader applications (Fig. 8-6, 8-7):

- Blister packages
- Foamed food containers
- Refrigerator and dishwasher door liners
- Auto interior panels
- Tub/ shower shells, which are later fiber reinforced
- Pickup truck bed liners
- Lighted signs made of acrylic or cellulose acetate butyrate (CAB)

Thermoformed Part Quality

The quality of a thermoformed plastic part depends on the type of mold used, the type of plastic used, and the method of thermoforming.

Molds. In a thermoformed part manufactured on a female mold, wall thickness tends to vary, becoming thinner as the plastic is drawn into the mold.

Fig. 8-6 Thermoformed containers. Courtesy of Brown Machine Division

This happens because the area of the plastic sheet available to form a part on a female mold is only a small percentage of the area of the formed part. Parts molded in a male mold may have the thin section on the opposite side. The dramatic variation in wall thickness can be reduced by using techniques such as plug-assisted molding. Molds for thermoforming must also have liberal radii (no sharp edges) to enable the plastic to conform to the mold easily.

Materials. The sheet plastic used for thermoforming has a major impact on the thermoforming process. Some plastic materials, like polycarbonate, have to be dried prior to thermoforming, because they are hygroscopic (absorb moisture) and moisture will weaken the plastic part. Amorphous plastics, such as ABS, styrene, and CAB, are excellent forming plastics because they maintain their heat and stretch well. Semicrystalline plastics, such as polyethylene and nylon, are more difficult to thermoform because they do not have a wide

Fig. 8-7 Thermoformed boat hull and liner. Courtesy of Brown Machine Division

processing temperature window. Styrene-based plastics, such as polyethylene, ABS, and polystyrene foam, are the most widely used thermoforming plastics.

Methods. The method or technique used to thermoform plastic parts affects the part quality. There is wider variation in part quality with manually produced thermoformed parts than with automatically produced parts. Thermoform molds that are heated produce a better match to the surface quality of the mold, because the heat is maintained for a longer time. Secondary operations, such as trimming, printing, and assembly, also affect part quality.

The Future of Thermoforming

Thermoforming will continue to be used primarily in the packaging markets. However, the use of thermoformed shells or exteriors that are subsequently reinforced with fiberglass will have a pronounced impact on the building and construction market (tubs/ showers and building panels), which has previously been served by sheet metal products.

Pressure Forming

Pressure forming is a variant of the thermoforming process. Thermoforming has the advantage of providing easy-to-manufacture parts on relatively simple and low-cost tooling, but the disadvantage is that it relies solely on vacuum or atmospheric pressure (14.7 psi) to form the part. As discussed above, the resulting plastic part shows significant variations in wall thickness and requires liberal radii to allow the plastic sheet to form around corners. This limits the appearance and the ultimate appeal of thermoformed products.

Pressure forming incorporates additional pressure, sometimes over 100 psi, making it easier to force the softened sheet to conform to the shape of the mold. This additional pressure, when properly used, allows for more intricate part designs, more uniform wall thickness, and sharper corner radii than in conventionally thermoformed parts. Some manufacturers of pressure-formed products are able to produce low-cost parts with complex shapes that rival the detail of injection molded parts. This detail allows some injection molders and part designers to use pressure forming for low-production or preproduction prototypes prior to tooling expensive injection molds.

Figure 8-8 depicts the basic pressure forming process. The process is identical to conventional thermoforming, except that higher pressure is introduced above the softened sheet, which forces the plastic to conform to the mold. As in thermoforming, plug assists may be used to facilitate the forming of more complex parts.

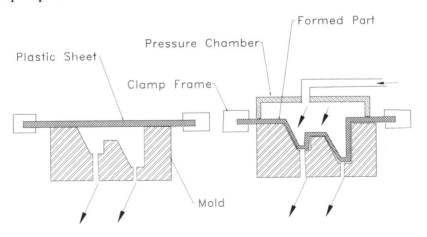

Fig. 8-8 Pressure forming

Resin Casting

Resin casting consists of pouring plastic resins that are fluid at or near room temperature into various types of molds, then allowing them to cure through the addition of heat. These thermoset resins undergo an irreversible reaction (see Chapter 1).

While resin casting can be automated to some degree, it has its greatest commercial value in prototyping and low-volume production. The greatest economic benefit occurs with unusual geometries that are too costly to machine and with annual volumes that are too low to justify the expense related to injection mold tooling. The tradeoffs, compared to high-volume methods such as injection molding, are lower tooling costs, higher unit costs, and a generally shorter tooling cycle. Figures 8-9 and 8-10 illustrate cast parts.

Most casting resins are low enough in viscosity to pour readily into molds without the need to force them in under pressure. In addition, the tempera-

Fig. 8-9 Cast products. Courtesy of Polymer Design Corporation

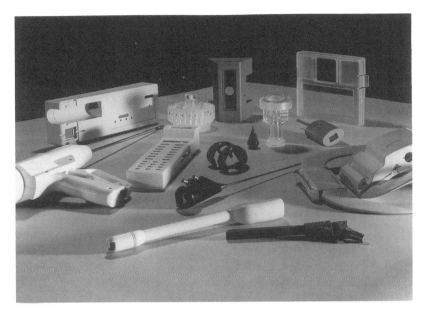

Fig. 8-10 Cast products. Courtesy of Polymer Design Corporation

tures required to effect an initial cure are typically 150 to 200 °F. These relatively mild processing conditions allow for the effective use of nontraditional mold media, such as room temperature vulcanizing (RTV) silicone. Other commonly used mold materials include steel, aluminum, epoxy, and polyurethane. These materials can be used in combination to take advantage of their individual attributes.

Ensuring that castings are cured without porosity requires the use of vacuum degassing. Coupled with other procedures, it offers the greatest flexibility in dealing with parts of different sizes. A prerequisite for vacuum degassing is that a material must remain liquid during the entire procedure, which typically means that a resin must remain liquid for 30 min or longer before hardening. Most formulations of this type need 10 to 12 hr to complete their initial cure. As a result, most molds are limited to one cycle per day, and typical production runs use multiple copies of molds.

Resin Casting Materials

Many manufacturing and design engineers assume that the performance of a particular material sample will be more or less the same for all formulations in that class of compounds. Nothing could be further from the truth. Each formulation has a unique set of processing attributes and final cured properties. In addition, the priorities of a prototyping application are quite different from those associated with the final production version. In light of the brisk pace of materials development, the best recommendations about what material to use for a given job come from suppliers and other users.

Epoxies are among the materials that have been used in commercial applications the longest. They continue to be at the forefront of materials technology, where they are used extensively as the backbone in resin composites. Epoxies are available in several forms, with the primary suppliers being CIBA Geigy, Dow Chemical, Shell, and Union Carbide. They are available in standard Bisphenol-A, Novalac, and Cycloaliaphatic grades, as well as others. These are combined with various curing agents and fillers to produce an almost infinite variety of compounds. For example, formulations are available to produce parts that will be:

- Clear
- Translucent
- Corrosion resistant
- Capable of withstanding 400 to 500 °F
- Compliant with FDA guidelines for wet or dry food contact
- Implantable in the human body
- Resistant to high voltages
- Suitable for use in high-strength flotation devices
- Microwave absorbing
- Microwave reflecting
- Autoclavable

By the creative combination of basic epoxies, curing agents, and various fillers, selected properties can be enhanced to accommodate a broad range of end-use applications.

Epoxies find their greatest use in applications where they must withstand a harsh environment. For the most part, they are highly crosslinked resins, which accounts for their ability to withstand extreme conditions. However, this also tends to make them somewhat brittle. Thoughtful designs will not incorporate thin sections that might be overly stressed. Successful parts designed with epoxies tend to be thick-walled "chunky" parts that use generous fillets and radii to eliminate stress concentrations.

Polyurethanes have as their greatest assets their toughness and abrasion resistance. While epoxies are very hard and brittle, polyurethanes are available in hardnesses ranging from Shore A-10 (soft rubber) to D-85 (hard plastic). In the hardness range of D-75 and softer, these materials are capable of handling standard drop tests and other more violent assaults. The softer the compound, the greater its ability to absorb energy without breaking.

Polyurethanes have several different forms; some common ones are polyester, polyether, polybutadiene, and aliphatic. They can be successfully blended with other resins, such as epoxies. This "alphabet soup" of chemistries allows the designer and custom processor to be creative and take advantage of the diversity and full capabilities of polyurethanes. Applications include:

- Instrument cases
- Decorative bezels
- Hand-held surgical devices
- Gaskets
- Cable strain reliefs
- Pulleys or slides
- Protective bumper strips and holders
- Fluidic devices
- High-voltage components
- Marine fittings (through hull, propeller shrouds)
- Lenses
- Electronics packaging
- Encapsulations

Polyurethanes do have limitations. They are generally not suitable in applications requiring elevated temperatures (above 200 °F), and they cannot withstand as broad a range of chemical exposures as epoxies can.

Silicones are the only inorganic compounds that are currently castable. The major U.S. suppliers are Dow Corning and General Electric. The RTV versions are most commonly used; however, other types of silicones, requiring heat and moisture to cure, are also castable. The major attributes of silicones are their ability to withstand temperatures as high as 600 °F and their uniform levels of flexibility and hardness at both high and low temperatures.

Resin Casting Tooling

The tooling for resin casting can be conventional metal molds (aluminum, steel, or brass) machined as one or more cavities, or quite unconventional RTV silicone molds, often used in combination with plastic cores. Other mold materials used include polyurethane, epoxy, and fiberglass. As in materials selection, the key is to ask, "What are you trying to accomplish?"

Silicone molds offer the advantages of high-quality finishes (release agents are commonly not required), casting in modest undercuts without side action, and reasonable tolerance control (±0.004 in./ in.). These molds are created by pouring RTV silicone rubber over patterns or models of the shapes of interest.

Patterns must be made of aluminum or some other suitable metal so they will not wear out or change shape with time. This allows processors to produce unlimited quantities of molds and parts, while maintaining the greatest control of dimensions and general part quality. Plastic or wood patterns may warp or distort with time, so their use is limited to prototyping applications.

While a sizing factor is normally applied in the creation of a pattern, the difference between the original and the subsequent castings can be insignificant in small to medium-size parts. This allows existing parts to be used as the basis for "tooling." The shape of the part, the required tolerances, and the appearance will determine how valuable an existing piece will be. Applications that require cores to define key details will not result in the same cost and time savings as those that can be made directly from a part.

A major advantage in using aluminum patterns and silicone molds in new product development is the relative ease of implementing changes. Many designs start in one form but evolve as they undergo the tests of manufacturability and exposure to the marketplace. Existing tooling is rarely scrapped unless changes are very radical.

In production, it is commonplace to run multiple quantities of silicone molds to increase the rate of manufacture of the part. Selected part details can be defined by metal or plastic cores (to reduce cost or to improve tolerance control) or by a machined secondary operation. If the number of machining operations becomes extensive, other mold materials (metal or plastic) should be considered.

Metal molds are capable of maintaining tolerance control of ±0.001 to 0.002 in./ in., with excellent flatness and parallelism of surfaces. As parts increase in size (above 300 in.^{2}), the unit cost of parts produced in metal molds becomes increasingly more favorable than the cost of the same part produced in a silicone mold. This is because silicone molds wear out and their cost is generally amortized into the part cost.

Finishes are limited to semigloss, because release agents are sprayed onto mold surfaces to ensure that parts can be removed properly. Draft may be required on parts that have an exceptionally deep draw. Careful production planning is required to match the production rate with the proper number of molds.

Epoxy/polyurethane molds provide some of the advantages of silicone and metal molds, but they also have some of the disadvantages. They offer the ability to make multiple copies of molds fairly inexpensively, as with silicone molds, but the quality of appearance is poorer than with any other mold option available.

The initial tooling is the metal pattern; then a suitable cast resin is poured over the pattern to create the mold. The pattern is not affected by the casting process, so it is possible to produce as many molds as required to support the production plan. Surface finishes are semigloss, because release agents are necessary. Tolerances are typically 0.002 to 0.004 in./ in. within a given mold. As additional molds are produced, each one is slightly different, and the larger the part, the more significant the variation becomes.

Epoxy/ polyurethane molds can wear out or even break with repeated heating and cooling. Replacement of molds may have to be considered in certain programs. Large parts (any single dimension over 40 in.) may be more economically produced by creating a wooden model than by producing a lay-up fiberglass mold. Parts of this type pose unique processing problems that must be addressed on an individual basis.

Resin Casting Design

Pouring free-flowing, slow-curing liquid resins into flexible silicone molds provides more opportunities for a designer than obstacles.

Modest undercuts can be cast without any side action. Some of the more flexible versions of silicone allow a 1.5 in. diameter part to be extracted from a 1.00 in. diameter opening. Contoured handled grips can be made without any parting lines, and doorknob-type features can be cast integral with a panel or tray. This particular advantage often allows a designer to combine two or more parts into one, potentially achieving meaningful cost savings. Extreme undercuts or openings can be accommodated by using multipart molds.

Replication of fine detail is another benefit of using silicone as a mold material. If the detail is on the pattern, it will appear on the parts, and this permits the casting of almost any finish. Typical cast finishes are high gloss, sand blasted, or textured. The key is that the original pattern must have exactly the same appearance as the finished part is supposed to have. Textures can be applied by painting the pattern or chemically etching the surface.

Tolerances achievable with silicone molds depend on a number of factors:

- How many parts are required from a mold?
- What is the mold material?
- What resin will run in the mold?
- Is it a one-part mold with a straight pull, or a two-part mold?

Another factor that can influence tolerance control is casting around a large metal component.

Wall thickness variations can be cast without any concern for sink marks or other blemishes, due to the use of slow-curing resins. Additionally, thick- or thin-walled sections can be cast almost without limitation. This and other design features are independent of the kind of mold material used. However, just because a detail can be created does not mean it will function as intended. The best recommendation is to consult an experienced supplier about the feasibility of a particular design detail.

Clear formulations are available in all classes of casting resins. These systems can be tinted with organic dyes to achieve custom tints at various levels of intensity. Some formulations darken or yellow with time, principally due to UV exposure, while others remain clear. The color shift has no effect on physical properties other than the slight change in visual appearance.

Custom colors can be achieved by blending various pigments into any resin formulation. This ability to cast color and finish into a part can save time and money, because there is no need to paint as a finishing procedure. Should the product be accidentally scratched, it will continue to show the desired color, thus maintaining a higher quality of appearance.

Some limitations in providing custom color should be noted. The base color of some casting resins is dark brown, making them inappropriate for applications requiring white parts. Also, because casting is a batch process (typically producing one to ten parts per day), there is wider part-to-part variation in color than with processes such as painting or injection molding, which produce many parts at one time.

A designer should not plan to use custom-colored cast parts immediately adjacent to molded or painted parts with a color-matching requirement. The results will be disappointing. In this situation, parts should be cast in complementary colors, so that slight color variations will not be noticed. If this is not possible, they should be painted with the matching parts.

Draft is never required on exterior surfaces, even with metal molds. Internal draft is necessary only in very deep draw applications where rigid cores are used.

Inserts and other foreign components are cast into parts routinely. It is often more efficient to add these components as a secondary operation than to cast them in place as part of the initial process. With proper technique, the inserted item is retained as securely as if it were cast in place.

Threaded metal inserts are available in self-tapping, expansion, and bonded-in styles. Metal threads are recommended over plastic when thread size of $\frac{1}{4}$ to 20, or smaller, is planned and repeated assembly and disassembly of the unit is anticipated.

Casting logos or other graphic details into a part lends a certain elegance to the appearance of an item. If desired, these can subsequently be filled in with a contrasting-color resin to make the graphics stand out.

Shielding against electromagnetic interference and radio frequency interference is best accomplished by painting a conductive coating (such as copper or nickel types) on surfaces of interest. Adding conductive metal flakes or fibers necessary to achieve a minimum level of shielding effectiveness increases the viscosity of the liquid resin so much that it is impossible to pour a high-quality casting.

Electrostatic discharge protection can be accomplished by adding specialty compounds, such as internal antistatic agents, to the resin.

Plastics Recycling

When it comes to waste and environmental issues, plastic products are singularly the most visible component of the waste stream. This is not because they are the largest-volume product discarded (Fig. 8-11), or because they cannot be recycled, but because there are only a limited number of aggressive and successful recycling efforts for plastic products. It is difficult to differentiate one plastic from another, and it is even more difficult to separate products manufactured with different plastics. Success must also be measured in terms of economics, because recycling efforts must be economically attractive if they are to proliferate.

Separating Plastics for Recycling. Figure 8-12 indicates three possibilities for recycling assembled parts that contain multiple plastics, or those that contain plastic plus other materials, such as metal or glass:

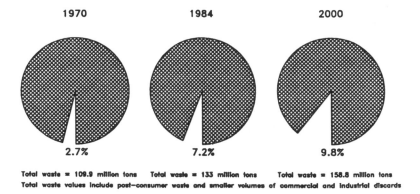

Total waste = 109.9 million tons   Total waste = 133 million tons   Total waste = 158.8 million tons
Total waste values include post—consumer waste and smaller volumes of commercial and industrial discards

Fig. 8-11 Plastics as a component of U.S. municipal solid waste. Courtesy of Franklin Associates, Ltd.; reprinted from "Waste Solutions," *Modern Plastics Magazine,* McGraw-Hill Publishing Co., April 1990

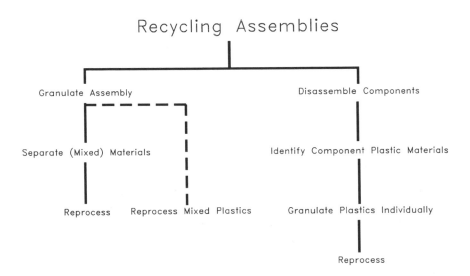

Fig. 8-12 Three possibilities for recycling assembled parts that contain multiple plastics and/ or plastic plus other materials, such as metal or glass

- Granulate the entire assembly. Once granulated, the chopped material can be separated with any of a wide variety of techniques, such as flotation or electrostatics.
- Granulate the entire assembly and reintroduce this granulate into another product, such as plastic "lumber" or a hybrid concrete formulation.
- Separate the materials, then granulate them individually and reintroduce them into their respective processes.

Figures 8-13 and 8-14 illustrate the use of polyethylene terephthalate (PET) in pop bottles and a typical flow sheet for reclaiming chopped PET. In this process, bailed PET is chopped via a granulator to reduce the particle size below 3/8 in. The chopped bottle can be processed to remove paper, fiber adhesives, and the high-density polyethylene (HDPE) base cup material. The starting aluminum content remains with the PET product after the label and base are removed. After drying, PET is cleaned electrostatically to remove over 99% of the aluminum. Treatment options include an aluminum clean-up step to further upgrade the conductor reject to over 90% aluminum. The electrostatically cleaned PET is then polished, reducing the aluminum content from 100 ppm to less than 10 ppm with less than 2% weight loss of PET flake.

Electrostatic separators are designed to process coarse particles and difficult-to-separate materials, such as:

- Chopped wire/plastic
- Precious metals/granulated electronic scrap
- Silver/melting furnace slag
- Chopped powder/Al_2O_3
- Chromium metal/alumina
- Raw materials for glass manufacturing
- Nonferrous metal
- Chopped plastic bottles (PET)

Electrostatic separation is based on the differences in shape, surface charge, or surface conductivity of the various particles, which typically range in size from 150-mesh to 0.5 in. (Fig. 8-15 to 8-17). Separators are available in several sizes, from pilot plant units capable of processing several hundred pounds per hour to large industrial units capable of processing up to 6 tons/hr. Suitably prepared materials should be mechanically liberated, dust-free, and surface-

Aluminum Cap   1g

PET   63g

LABEL
& ADHESIVE   5g

HDPE BASE CUP
& ADHESIVE   22g

Fig. 8-13 Materials in a typical two-liter beverage bottle. Courtesy of the Plastic Bottle Institute

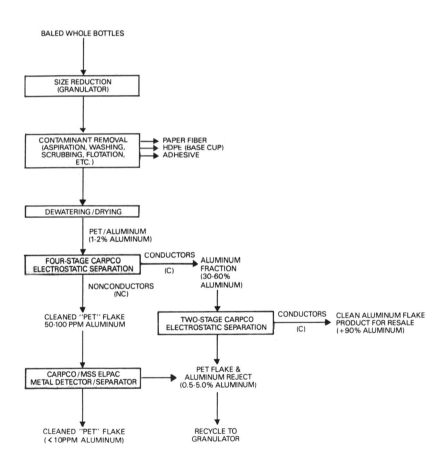

Fig. 8-14 Typical flow sheet for reclaiming chopped PET. Courtesy of Carpco, Inc.

Fig. 8-15 Machine designed to separate aluminum and other nonferrous contaminants from recyclable plastics. Courtesy of Carpco, Inc.

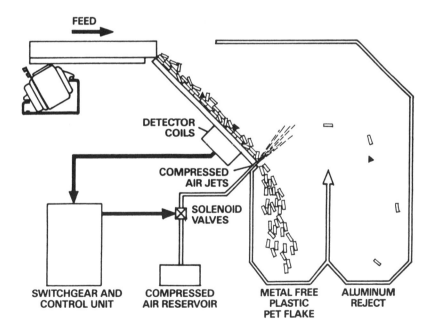

FEED

DETECTOR
COILS

COMPRESSED
AIR JETS

SOLENOID
VALVES

SWITCHGEAR AND
CONTROL UNIT

COMPRESSED
AIR RESERVOIR

METAL FREE
PLASTIC
PET FLAKE

ALUMINUM
REJECT

Fig. 8-16 Detail of metal removal from recyclable plastics. Courtesy of Carpco, Inc.

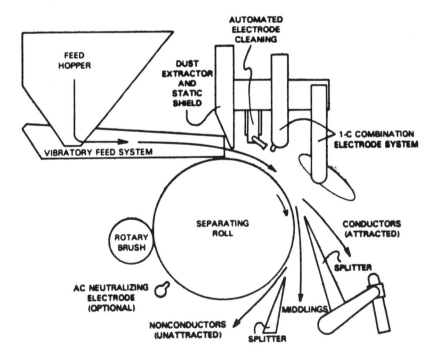

Fig. 8-17 Diagram of features in an electrostatic metal separator. Courtesy of Carpco, Inc.

dry. Separation can be improved for some materials at elevated feed temperatures.

Figure 8-18 is a flowchart for electrostatic separation of chopped wire tailings. Bailed scrap wire is coarse chopped (to 5 in.), then granulated to below ¼ in. Paper and fluff (fiber) released during granulation are removed through aspiration. The granulated wire is then processed through an air/ gravity table system, which removes more paper and fluff and produces a primary metal product. The remaining chopped plastic (PVC or polyethylene) typically contains 4 to 10% metallic components in the form of a flaky aluminum and fine "hair" copper wire, along with residual paper and fiber contaminants.

The chopped insulation is screened and aspirated again to remove additional paper and fluff, and the dust-free PVC or polyethylene insulation is cleaned electrostatically to remove aluminum and/ or copper contaminants. The final plastic product is typically less than 0.5% metallic and is suitable for recycling. The metallic concentrate from the electrostatic separator, 50 to 70% metallic, is then easily upgraded via a destoner to form a secondary metal product suitable for market. Aluminum flake remaining with the destoner light fraction can be cleaned with an optional electrostatic separator.

The Economics of Plastics Recycling. Morality and ethics have often been the focus of attempts to persuade individuals to recycle, but it is obvious that the need to recycle has gone well beyond them (Tables 8-1, 8-2). The volume of discarded materials in the United States alone is staggering. Landfills are filling up and closing, and new landfills are not being created quickly enough to absorb the continuous increase in solid waste. Common sense suggests that the solution to the solid waste disposal problem is not to have solid waste in the first place!

Plastic products and materials eventually degrade, but at a slow rate, maybe after 5 to 5,000 years. Even plastics that degrade in a relatively short period require sunlight and moisture to facilitate the degradation process. Most of the solid waste in landfills lies below several feet of dirt, so even plastic products designed to be biodegradable often have no chance to degrade. Furthermore, solid waste degradation may be detrimental to the environment if the degraded components find their way into the water supply.

Every successful recycling effort has been based on sound business and market planning. A quick look at the success story of aluminum can recycling reveals what is required for plastics recycling to be economically viable:

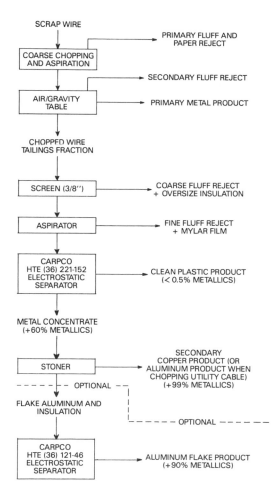

Fig. 8-18 Typical flow sheet for electrostatic processing of chopped wire tailings. Courtesy of Carpco, Inc.

Table 8-1 Recycling growth estimate

	1989 %	1989 Million tons	1994 %	1994 Million tons	Average annual growth rate, %
Glass	20	0.75	40	1.7	17.8
Metal	19.5	2.6	25.9	4	9
Paper	25	12	28.7	15.4	5.1
Paperboard	32.3	11	37.7	13.2	3.6
Plastics	5	0.1	13	0.4	31

Source: Business Communications Co., Norwalk, CT, reprinted from "Waste Solutions," *Modern Plastics Magazine*, McGraw-Hill Publishing Co., New York, April 1990.

Table 8-2 Reclaiming plastic from consumer waste

Material	Annual discards million ton	Consumer waste recovery, million ton	Recovery rate, %
Aluminum	2.1	0.6	28.6
Paper and paperboard	62.3	12.9	20.7
Glass	13.9	1.0	7.2
Rubber and leather	3.4	0.1	3.0
Iron and steel	11.3	0.3	2.7
Plastics	9.7	0.1	1.0
Total	100.6	14.4	14.3

Source: Franklin Associates Ltd., Prairie Village, KS, reprinted from "Waste Solutions," *Modern Plastics Magazine*, McGraw-Hill Publishing Co., New York, April 1990.

- Aluminum cans are easily identifiable. Not only is "Please Recycle" written on the can, but the consumer simply has to squeeze the thin can to verify that it is made of aluminum.
- Successful recycling of aluminum cans occurs in areas where there is a deposit or refund incentive. The success of the deposit can be appreciated when reviewing statistics about the low percentage of aluminum cans recycled in areas that offer no refund.
- Bulk separation of aluminum cans is also relatively easy, because the only similar product is the steel can, which can be magnetically separated.

All plastics can be recycled. Plastics that cannot be remelted and remolded may be granulated and used as filler materials in other plastic products, asphalt, or concrete. Plastics reusability is limited only by the imagination.

9

Ancillary Equipment

Ancillary equipment is used to support primary plastics processing equipment. Typical ancillary equipment includes:

- Plenum hoppers
- Material loaders
- Material dryers
- Magnet systems
- Mixing systems
- Robots and part pickers
- Conveyors
- Granulators
- Mold temperature controllers

While much has been written about primary processing of plastic materials to create plastic parts, very little has been written about ancillary equipment. Similarly, when manufacturers evaluate primary processing equipment for potential purchase, they usually conduct a thorough and extensive study; however, when it comes to purchasing ancillary equipment, too often few or no studies are conducted.

Fig. 9-1 Drying system

Drying

Drying Equipment

Most plastics drying systems consist of three main pieces of equipment:

- Plenum hopper
- Desiccant dryer
- Hot air dryer for removing surface moisture of nonhygroscopic materials

The hopper and dryer work together to provide the proper drying environment and residence time to dry plastic pellets prior to processing. Figures 9-1 to 9-3 illustrate a totally closed-loop, valveless drying system that has been

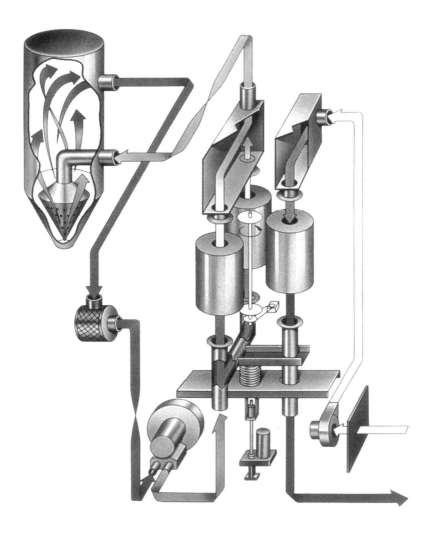

Fig. 9-2 Detail of a drying system. Courtesy of Thoreson-McCosh, Inc.

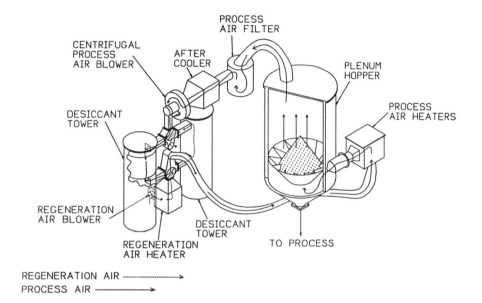

Fig. 9-3 Schematic of a drying system

engineered to efficiently dry plastic pellets and to keep moisture, dirt, and other contaminants away from the material. Its components are:

- *Plenum hopper:* used to hold the plastic materials in a controlled environment for proper drying. The hopper may be mounted either directly on the throat of the plastics processing machine or on the floor adjacent to the processing machine.

- *Diffuser cone:* used to allow the hot dry air to be spread uniformly up through the plastic pellets

- *"To process" air:* The hot dry air from the dryer enters the hopper in the lower part of the plenum hopper within the diffuser cone. It is important that the air temperature be measured at this point and controlled to the material drying specification.

- *"From process" air:* The warm moist air exits the plenum hopper from the top. From this point, the air is reintroduced into the dryer to have the moisture content reduced and the temperature re-established. The air from the hopper is recirculated, because although it has picked up moisture, it is still much dryer than the ambient air, typically 0 °F dew point.

- *Filter:* As the air moves through the dryer/ hopper system, it tends to pick up fines (small particles of plastic). The fines must be removed from the air to allow maximum drying efficiency and to prevent damage to the molecular sieves.

- *Blower:* Warm moist air exiting from the plenum hopper is directed through the desiccant bed(s) and heaters prior to being reintroduced into the hopper.

- *Desiccant beds:* These canisters contain a desiccant ceramic material called molecular sieves. The sieves allow air to pass through them, but they trap moisture. Molecular sieves should be inspected periodically, and they should be replaced if they are found to be contaminated.

- *Process air heater:* The freshly dried air is now heated to the desired temperature by passing through this heater cartridge.

- *Regeneration of desiccant bed(s):* The desiccant bed(s) are periodically rotated to a regeneration station where hot air (about 550 °F) is forced through them. Any moisture present is allowed to exit outside the process loop.

- *Regeneration heaters:* used to raise a separate airstream (not the process air) to 550 °F for the desiccant bed regeneration station

The Necessity of Removing Moisture

Many thermoplastic materials are hygroscopic, meaning they absorb moisture. The amount and rate of moisture absorption varies, depending on the base polymer and additives, but this moisture must be removed prior to plastics processing. If the moisture is not sufficiently removed, the processing temperature will cause the water to turn to steam, resulting in problems during processing, cosmetic defects on the part, or voids in the part that could be sites for product failure. In some materials (e.g., polycarbonate), any moisture present will result in hydrolysis when the material is heated, and a breakdown of the molecular structure will occur. This will result in an irreversible loss of physical properties, and the material will have to be scrapped.

The Equilibrium Moisture Curve

The drying of any granular solid material requires that two fundamental processes take place:

- *Evaporation of surface moisture on the granule:* This occurs by providing an environment of lower partial pressure than the vapor pressure of the surface moisture present on the granule.
- *Diffusion:* After the surface moisture has been removed, a moisture gradient is established within the granule, and moisture migrates to the surface of the granule to balance the internal moisture concentration.

The result of the combined effect of evaporation and diffusion is shown graphically in an "equilibrium moisture curve" (Fig. 9-4). Such a graph indicates the moisture content, by weight, of a specific thermoplastic material (in this case polycarbonate) versus the time the material is exposed to the drying environment. The drying environment must also be specified; in this case it was 250 °F and –20 °F dew point. A horizontal line is often shown on the equilibrium curve to indicate the acceptable moisture level for molding or extruding the material with no loss of physical or appearance properties as a result of the residual moisture. This value is stated in the processing and design guides provided by most plastic material manufacturers.

The size and shape of the granules affect how the material will respond to drying conditions, in terms of drying efficiency and effectiveness. Smaller granule size contributes to more rapid drying, as does any shape of the granule that increases its surface area without increasing the distance from the center to the surface. Size and shape are normally not significant when drying virgin materials, but they can be factors when drying regrind, particularly if large particles (greater than 0.060 in. from center to surface) are present.

Another factor that influences the drying rate is the pressure within the drying vessel. Lower pressure increases the drying rate, so that there is less time to bring the material down to an acceptable moisture content. Vacuum dryers have not been used extensively in the plastics industry due to their cost, their complexity, and the fact that they are not very adaptable to continuous throughput processing. Today, the majority of hygroscopic thermoplastic materials are essentially dried at atmospheric pressure.

Referring again to the equilibrium moisture curve (Fig. 9-4), note that the moisture content of a particular material is a function of the temperature and dew point of the drying process stream, as well as the length of time the

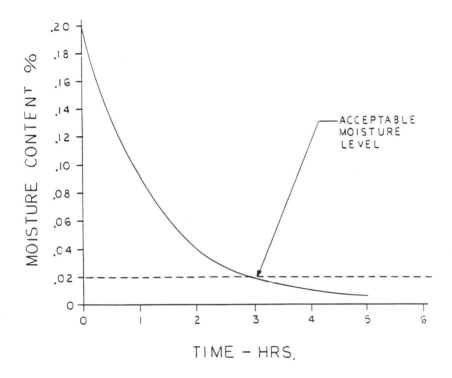

Fig. 9-4 Equilibrium moisture curve for polycarbonate under drying conditions of 250 °F and –20 °F dew point

material is exposed to the stream. It is important to keep in mind the basic response of the material to these three principal factors in drying performance. Nearly all drying problems can be traced to lack of control of one or more of these factors.

The Hopper

Having established what must be done to the individual plastic granule to bring its moisture level down to a point that will eliminate problems, one must then apply this treatment to the total quantity of granules that the process requires. A drying vessel or hopper must be selected that is of adequate size and designed in such a manner that all granules flowing through the hopper are exposed uniformly to the drying process stream. This sounds rather elementary, but careful attention must be paid to the design of the hopper and diffuser combination, to ensure uniform distribution of the process stream up through the hopper and uniform downward progression of the material being dried. Channeling of air or nonuniform flow of the material can readily affect the time/ temperature/ dew point relationship required to properly dry the material.

It is difficult for the purchaser of a drying hopper to determine from appearance just how well the hopper will perform in terms of air flow and material distribution. Evaluating hopper performance involves suspending elaborate thermocouple networks within the hopper to determine the heat progression through the hopper, and using differently colored materials to determine the time distribution of material as it flows down through the hopper.

In Fig. 9-5 the progression of hot air up through the hopper is plotted across the hopper section. As indicated, the progression of the temperature change and air flow is uniform across the hopper section. Some decrease in temperature is evident near the surface of the hopper, due to the loss of heat at the surface. Insulation of the hopper, which is normally available as an option, will minimize heat loss through the hopper surface and result in less temperature dropoff at the outer edges of the temperature curve.

Figure 9-6 illustrates the residence time distribution of material in a hopper that is loaded 50% full with material of one color and 50% full with the same material of a different color. The hopper is then emptied, preferably with an augur device, which creates a condition similar to the feeding of an injection molding machine or an extruder. If the hopper is properly designed, the transition from one color to the other takes place over a relatively short time span, as the graph's steep curve illustrates. Ideally, the transition would take place at once, but practically speaking the transition will take place over a period of time approximately ±15% of the nominal residence time. The significance of the rapid and complete color transition goes back to the fundamental time/ temperature/ dew point relationship: it shows that all the material granules have been exposed for the desired residence time.

While it is not expected that plastics processors will run the above-mentioned tests to determine the performance of the drying hopper, it is important that they understand the function of the hopper in a good drying system, so they can be certain the equipment will perform properly.

The Dryer and Drying Requirements

The dryer's function in the system is to supply an adequate quantity of air at the correct temperature and dew point. In this respect, the dryer can be

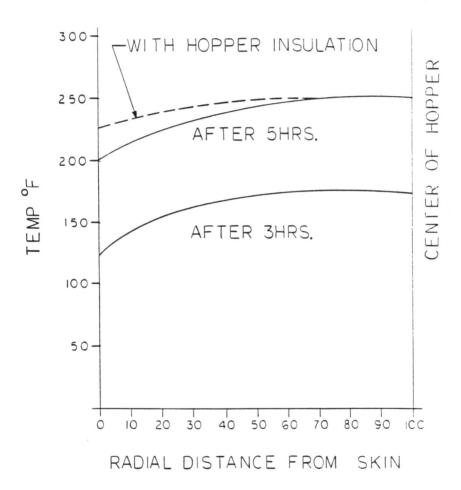

Fig. 9-5 Hopper temperature gradient

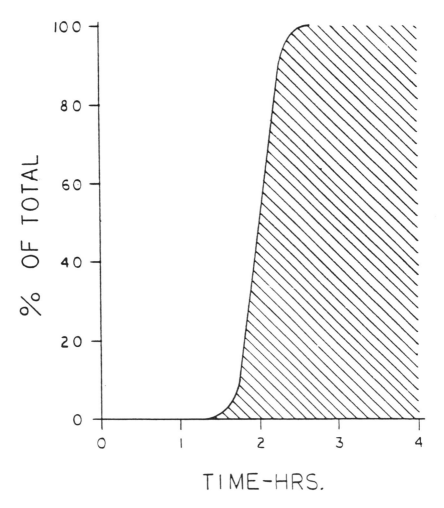

Fig. 9-6 Material residence time distribution

considered an air conditioner, a source of air that is properly conditioned in terms of time, temperature, and dew point. It is important to recognize that the actual material drying is done in the hopper.

In selecting a dryer, the important factors to consider are:

- The cubic feet per minute of air discharged into a pressure drop equivalent to the full drying hopper. Some dryers being sold today are rated in terms of their free air delivery (i.e., their discharge into the room). This rating means little or nothing in terms of the unit's ability to move the drying air through the hopper filled with plastic material.
- The ducting between the hopper and the dryer
- The dew point of the discharge air stream
- The maximum temperature regulation capability of the dryer at the specified cubic feet per minute

All desiccant dryers being marketed today for drying plastic materials use the same adsorption material: molecular sieves. The capability of this material for adsorbing moisture and the energy required to remove the moisture from the sieves during the regeneration are fixed. A purchaser of a desiccant dryer should be aware of this fact and determine the weight of the sieves in the dryer. The air flow capability of a dryer is nearly proportional to the quantity of the sieves in the dryer.

With all this emphasis on the air flow capability of the dryer, it is important to establish the air flow required to adequately dry a given quantity of material. The air stream flowing up through the hopper provides two essential functions in the drying process:

- It establishes the dry environment around the granules to promote evaporation.
- It acts as the medium of heat transfer required to bring the temperature of the material up to the prescribed drying temperature, in order to speed up both the evaporation and diffusion processes.

The second function determines how many cubic feet per minute of air are required to achieve proper drying. In determining the required air flow, let's first assume that we are going to dry 100 lb/hr of polycarbonate with 0.35% water content (recommended drying temperature 250 °F). We know that the specific heat of the resin is 0.35 Btu/lb °F, the specific heat of the air is 0.24

Btu/lb °F, the change in air temperature across the hopper is 130 °F, and the thermal efficiency of the drying hopper and its associated duct work is approximately 50%. In other words, only half of the heat being discharged from the dryer stays in the material being dried and increases its temperature. The remaining 50% is lost from the surface of the drying hopper, lost from the ducts, or remains in the air stream when it leaves the hopper.

Equation 1 shows how to determine Q, the heat required to raise 100 lb/hr of polycarbonate to a prescribed drying temperature:

Heat required to raise 100 lb/hour of material to 250 °F:

$$Q = \left(W_p\, C_p\, \Delta_{tp} + W_w\, r + W_w\, C_w\, \Delta_{tw}\right) e$$

$$Q = (100 \text{ lb/ hour} \times .35 \text{ BTU /lb } °F \times 180 \text{ }°F$$
$$+ .35 \text{ } 970 \text{ BTU/ lb}$$
$$+ .35 \times 1 \text{ BTU/ hour } °F \times 180 \text{ }°F) \frac{1}{.5}$$
$$= 13{,}400 \text{ BTU/hour}$$

where W is weight, C is specific heat, and Δt is temperature change.

Equation 2 shows how to determine W_a, the air flow (in ft^3/min) required to provide Q:

Air flow (cu ft/minute) required to provide Q:

$$W_a = \frac{Q}{C_a\, \Delta_{ta}}$$
$$W_a = \frac{13{,}400 \text{ BTU/hour}}{.24 \text{ BTU/ lb } °F \text{ } 130 \text{ }°F}$$
$$W_a = 430 \text{ lb/hour}$$

To convert lb/hour to cu ft/minute:

$$430 \text{ lb/hour} \times 14 \text{ cu ft/lb} \times 1 \text{ hour/}$$
$$60 \text{ minute} = 100 \text{ cu ft/minute}$$

Key:

W = weight, C = specific heat, Δ = change

Then lb/ h can be converted to ft³/ min:

To convert lb/ h to cu. ft./ min.

$$\left(430\frac{lb}{hr}\right)\left(14\frac{ft^3}{lb}\right)\left(\frac{hr}{60min}\right) = 100\ ft^3/min$$

The result, 100 ft³/ min, means that the ratio of air to material is 1 ft³/ min/ hr. The required air flow rate is in general agreement with that established by other investigators (Miles Report on Polycarbonate Drying TSA-70-015). It represents a system that will dry virtually any material on any given day.

Aside from the primary drying requirements mentioned above, the dryer should be small, efficient, reliable, and as easy to use as possible. In terms of efficiency, it is important to consider that the requirements for moisture removal vary considerably (by a factor of 6 to 1) with variations in the moisture content and ambient conditions of the materials being dried.

Any dryer control system capable of making significant gains in overall efficiency must be able to sense changes in moisture load requirements, then compensate by adjusting the amount of energy used to regenerate the molecular sieve. This system is in contrast to the conventional method of regeneration, which is based on a fixed time cycle or rate of bed rotation. One method of matching the amount of regeneration energy to the drying load is to control the time that the regeneration heat is applied. This can be accomplished by monitoring the amount of moisture remaining in the bed and shutting the heat off when the moisture is reduced to a level that indicates that regeneration is complete.

A great deal has been said about the discharge dew point of desiccant dryers. Essentially the discharge dew point is a function of how well the beds are regenerated. Again, from a practical standpoint, it has been demonstrated that while drying rates increase as dew points are lowered, a point of diminishing return is reached at approximately –20 °F, which is adequate for all materials being dried today.

Considerable controversy has been raised recently regarding the variation in dew point when a twin bed dryer is switched from one bed to another. Proponents of rotating bed dryers have stated that the dew point of a twin tower dryer fluctuates as much as 40 °F during bed changeover. Such statements are in error and could only apply to dryers that were grossly undersized, so that the breakthrough occurred well in advance of the changeover to the newly regenerated bed. A properly designed twin tower dryer produces an essentially uniform discharge dew point and has the same material drying capability as any other type of dryer.

The dryer must have adequate heating capability to bring the temperature of the process stream up to the drying temperature prescribed by the material supplier. Except for polysulfone and PET, most of the hygroscopic thermoplastics can be dried with a dryer that is capable of 300 °F. Temperature regulation of the drying stream should be adequate to prevent any problems with material degradation due to excessive temperature overshoot.

To summarize, the conditions required for good drying of hygroscopic materials are:

- Adequate residence time (3 to 4 hr)
- Good air flow distribution through the hopper (±15 °F temperature variation across the hopper)
- Good residence time distribution (±15% of nominal residence time)
- Adequate quantity of drying air through the hopper (1 ft^3/ min/ hr being dried)
- Adequate drying temperature for the quantity of drying air through the hopper (250 °F minimum)
- Adequate dew point for the quantity of drying air through the hopper (–10 °F or lower)

While the topic of drying hygroscopic thermoplastic materials appears rather simple, it has nonetheless been a source of much frustration and considerable expense to processors who have not done their homework and obtained adequate equipment. In the winter, when ambient dew points of 0 to 20 °F are common, one may be lulled into believing that there will be no problems with drying materials. This belief is often shattered when the hot, humid days of July and August arrive and dew points of 70 °F or better are not uncommon. It does not take many days of operating with high reject rates to become convinced that the price of an adequate drying system is justifiable.

Preventive Maintenance of Dryers and Hoppers

One of the most unfortunate truths within the plastics processing industry is that processors often have material drying problems. However, the majority of these problems are self-inflicted and easily prevented. Some of the key preventive maintenance operations are to:

- Inspect and clean all air filters on the dryer at each materials change or every 24 hr, whichever occurs first.
- Inspect the desiccant (molecular sieves) every 6 months and replace as required. Material fines often enter the molecular sieve bed and decrease their effectiveness.
- Inspect the process air heater element and the regenerating air heater elements every 6 months, and calibrate as required.
- Strip and clean the hopper and diffuser cone (using a vacuum, not an air hose) prior to every materials change. Remove all fines and clean all screen holes.
- Inspect all seals and gaskets on the hopper and replace as required. To work properly, the hopper must be a controlled environment sealed from any outside moisture and contamination.
- Enclose the hopper in an insulating blanket, if it is not already insulated. This will prevent heat loss and create a more energy-efficient system.

Trends in Material Dryers

One of the areas of concern with plastic material dryers is efficiency, both energy efficiency and the rate at which material is dried. Two innovative types of dryers are under development.

Natural gas dryers have already been marketed, but with only limited success. In concept, they are somewhat similar to gas clothes dryers; in mechanics, they are virtually identical to electric dryers, with the exception that the heat is generated by burning gas. The heat generated in the combustion of gas (or any hydrocarbon fuel) produces water, so gas heating of the process stream must be done with heat exchangers in order to maintain low dew points.

Gas dryers for plastics have not been widely accepted because of the perceptions that gas-fired equipment is not movable and that it may not be as safe as electricity. Both of these perceptions are unfounded, and as future energy costs rise, natural gas dryers may see more use in the plastics industry.

Microwave dryers are under development in Japan and have been visible at various plastics expositions over the past five years. They use the same technology to dry plastic as is used to pop popcorn in your home microwave. The microwave energy excites the polar water molecules, creating friction and heat. When enough heat is generated, the water exits the plastic pellets. The future of microwave drying is bright, mainly because a plastic that requires 4 hr of residence time in a conventional hopper/dryer may require less than 1 hr in a microwave dryer, which will allow for more rapid material changes. The two disadvantages of microwave drying are its current high cost and the fact that most of the prototype equipment has been batch-oriented with a limited capacity.

Material Loaders

Plastic material loaders are used by all plastics processors. They allow material to be moved from a material storage container or drying hopper directly to the plastics processing equipment without being exposed to moisture and contamination. The alternative to using a material loader is to load hoppers manually, which is not a wise decision because it:

- Is messy
- Exposes dry material to moisture (unless the material will be dried on the machine after loading)
- Requires the operator to perform risky tasks (lifting, climbing equipment, and pouring material)

Some processors have attempted to save money by building simple venturi rigs, loading wands outfitted with a highly directed stream of compressed air. The rapidly moving air stream creates a low-pressure vacuum effect that forces the plastic pellets into the hopper. Although inexpensive, venturi rigs are:

- Noisy
- Messy (they move large volumes of air)
- Problematic because they introduce moisture to the hopper
- Difficult or impossible to control
- Expensive to operate due to the large quantity of compressed air required

Automatic loaders are convenient as well as safe, and if properly maintained they will last for decades. The feature that is critical to plastics processing is that automatic loaders allow processors to determine loading sequences and to control environmental factors, such as by establishing a completely closed system.

All loader systems have the same basic components:

- A *motor* creates a low-pressure environment within a small chamber. This motor is often situated as an integral part of the loader, but in larger production facilities a centralized motor services several pieces of processing equipment.
- A *filter system,* usually in the form of a flat fabric filter, helps keep fines from entering the motor and manufacturing area.
- A *paddle switch* hangs into the hopper and senses when more material should be loaded.
- A *control system* provides the plastics processor with some latitude to set loading times and sequences.

There are four basic classifications of plastics loaders:

- A *standard hopper loader* (Fig. 9-7) is a self-contained unit that is attached to each machine. Pellets are usually loaded into a drying hopper.
- A *central hopper loader* (Fig. 9-8) has a central vacuum system that services several processing machines.
- A *mini hopper loader* (Fig. 9-9) loads plastic from a drying hopper that sits on the floor, leaving only a small quantity of plastic above the throat of the machine.
- An *additive feeder* (Fig. 9-10), although not a loader, feeds additives (e.g., colorants) at a prescribed rate.

Granulators

Granulators, like many pieces of ancillary equipment, may not receive the same degree of scrutiny as primary processing equipment when it is time to make a purchase. Many top plastics processors have technicians trained to operate the most sophisticated process-controlled molding equipment, but these same technicians may know little about the use and care of plastics granulators.

Plastics granulators were originally designed, 70 years ago, with two jobs in mind:

- Pulverizing polymer directly out of the polymerization kettle, to provide a configuration more usable by processors of phenolic and early thermoplastics

Fig. 9-7 Standard hopper loader. Courtesy of Thoreson-McCosh, Inc.

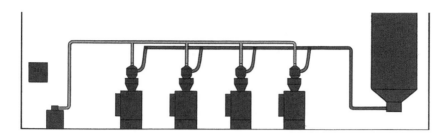

Fig. 9-8 Central hopper loader. Courtesy of Thoreson-McCosh, Inc.

Fig. 9-9 Mini hopper loader. Courtesy of Thoreson-McCosh, Inc.

Fig. 9-10 Additive feeder. Courtesy of Thoreson-McCosh, Inc.

- Granulating thermoplastic sprues, runners, and scrap parts on the production floor, allowing the plastic material to be mixed with virgin plastic and reintroduced into the process

Both original applications of granulators are still used today; however, the demands put on granulators have become more specialized. Plastics processors are attempting to reduce or eliminate nonproductive material, such as sprues, runners, and bad parts, but all still exist. Some processors would like to recover the otherwise wasted material generated from process equipment start-ups and shutdowns. Some want granulators next to processing equipment, while others collect material and regrind it at a more central point off the

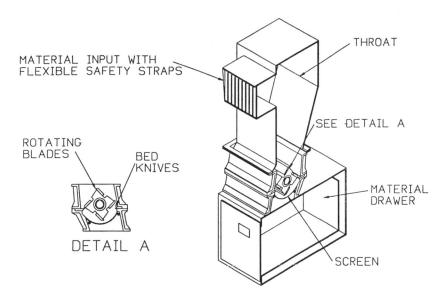

MATERIAL INPUT WITH
FLEXIBLE SAFETY STRAPS

THROAT

SEE DETAIL A

ROTATING
BLADES

BED
KNIVES

MATERIAL
DRAWER

DETAIL A

SCREEN

Fig. 9-11 Granulator

production floor. Figure 9-11 depicts a typical granulator that could be used beside the plastics processing equipment.

Regardless of the use and application of the granulator, some common features should be considered:

- *Safety:* All granulators should have safety interlocks to reduce the risk of injury to the operator. While most of the safety features are in the area of the cutting chamber, the working portion of the throat should be designed such that it will not allow the operator's arm to pass inside.

- *Noise:* The design of the cutting chamber, the amount of sound-deadening material, and the type and shape of the parts to be granulated all affect the noise that is generated. Operation below 85 dB is a must; operation below 80 dB is desirable.

- *Feed:* Depending on the size and type of items to be granulated, a processor may want a manually loaded granulator, an augur feed granulator, or a granulator that is linked with a custom materials handling system to facilitate loading.

- *Cutting blades:* Granulator cutting blades are made from high-carbon, high-chrome steel. They are usually classified as either bed knives, which are fixed, or rotating knives, which move. The relationship of the bed knives to the rotating knives is usually angular, to allow for a scissor-like cut.

- *Screen:* The screen or mesh plate resides just below the cutting area. The holes in this screen or plate must be carefully selected. If the material to be granulated is too small, as with skinny sprues, it will pass through the cutting chamber without being cut.

- *Granulate:* Material that has been granulated can be manually emptied via a drawer (Fig. 9-11), but chutes and automatic unloading devices facilitate the use and unloading of the granulator in high-production situations.

- *Ease of disassembly:* Granulators require frequent disassembly and assembly for materials changeovers and maintenance. This fact should be carefully considered when selecting a new unit. If the unit does not easily come apart, several hours of labor will be lost during every materials change.

- *Preventive care:* Granulators should be thoroughly cleaned during every materials change. The cutting chamber should be carefully vacuumed, and the screen and blades should be inspected for wear and proper alignment. When blades are worn they can be resharpened, but they may then have to be realigned. Remember that when blades wear, tramp metal is deposited into the granulate and must be removed with a magnet prior to reprocessing.

Figures 9-12 and 9-13 illustrate the cutting chambers of three- and four-blade granulators, respectively. In both illustrations, item 10 is the screen/ mesh plate.

Mold Temperature Controllers

Controlling the temperature of a mold or die is critical to producing quality plastic parts. As molten plastic is forced into or through a mold or die, the thermal energy generated during the melt phase of the process needs to be dissipated in a controlled manner. The mold/ die has to be properly designed

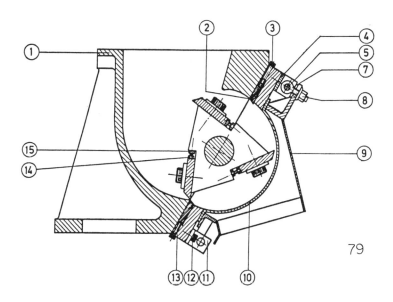

Fig. 9-12 Three-blade granulator. Courtesy of Rapid Granulator, Inc.

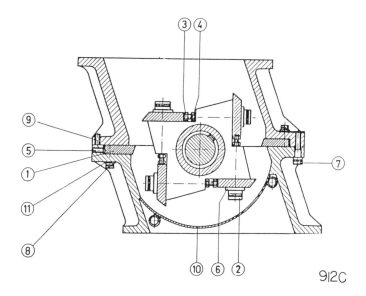

Fig. 9-13 Four-blade granulator. Courtesy of Rapid Granulator, Inc.

and machined, and the medium selected to accomplish the heat transfer must be effective.

The medium used to control mold and die temperature depends on the temperature required:

Medium	Temperature required, °F
Ethylene glycol and water	−20 to 70
Water	50-210
Electricity (cartridge heaters)	37-200
Oil	200 to 450+

For thermoplastics, it is important to remember that regardless of the method of temperature control, the mold is being cooled. Even if the mold requires a temperature of 350 °F, the temperature of the mold is lower than the temperature of the plastic melt. Another way to state this is that if heat is being transferred from the plastic to the mold, the mold is cooling the plastic. If the heat transfer is the reverse, the mold is heating the plastic.

Some prototype and low-production tooling may have no provisions for heating or cooling. A heater unit and pump are required for liquid media, while temperature controllers and temperature sensors are required for electric heaters.

Circulating Water Temperature Controllers

For molds or dies that require water for heating or cooling, there are several considerations in selecting a circulating water temperature controller:

- Size (gal/min)
- Capacity (Btu/hr)
- Open- or closed-loop
- Portable or built into the plastics processing equipment
- Single or multiple zones

Figure 9-14 illustrates the major components of a circulating water temperature controller. Some key features should be considered in the selection, operation, and maintenance of the system:

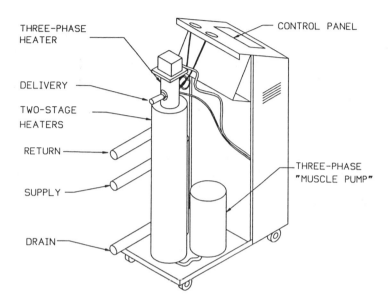

THREE-PHASE
HEATER

CONTROL PANEL

DELIVERY

TWO-STAGE
HEATERS

RETURN

THREE-PHASE
"MUSCLE PUMP"

SUPPLY

DRAIN

Fig. 9-14 Circulating water temperature controller

- *A low-pressure valve* to prevent the pump and system from operating if there is low water pressure in the system, thus reducing the chance of pump damage
- *Solid state temperature controller(s)* for accurate and precise temperature setting and control
- *High-temperature safety* to reduce the risk of temperature spikes that may cause processing problems or operator injury
- *A venting option* to allow easy removal of entrapped air that may cause cavitation
- *Easy access to components* for maintenance

• *Portability* to allow proper positioning next to the processing equipment

Most processors of injection molded and blow molded parts prefer a water control system that has multiple zones. These systems are basically "ganged" water control units that allow the processor to keep mold halves at different temperatures without using multiples of single units. Some equipment manufacturers integrate a water control system directly into the primary plastics processing machine for convenience and to reduce clutter around the equipment.

Open/Closed Circuits. Figures 9-15 and 9-16 illustrate the differences between open- and closed-loop water control systems. The more common system is open-loop, or direct (Fig. 9-15). This system allows for direct introduction of cold water into the controlling water circuit, while instantaneously

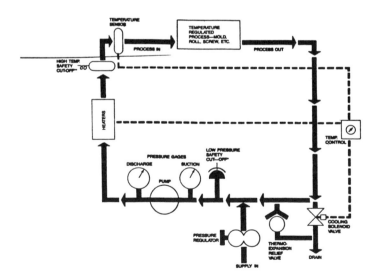

Fig. 9-15 Open-loop (direct) cooling circuit. Courtesy of Mokon

expelling hot water to the drain for purposes of cooling and overall temperature control.

The closed-loop, or indirect, cooling system (Fig. 9-16) isolates the controlling water circuit from the cooling water circuit via a shell and tube heat exchanger. This system is often used in cases of bad water supply or where glycol solutions are used for the controlling circuit.

Negative Pressure Systems. Where circulating water is used for temperature control of molds, rolls, jacketed vessels, and similar equipment, a negative-pressure-based circulating water system provides inexpensive insurance against interruption of production runs due to water leaks (Fig. 9-17). These systems are used with standard water temperature controllers. Water flow is controlled by means of a regulated vacuum pump that "pulls" rather than "pushes" the water through the process. Air is therefore drawn into, not forced out of, leaking areas around faulty O-rings, holes in pipes, and cracks in molds.

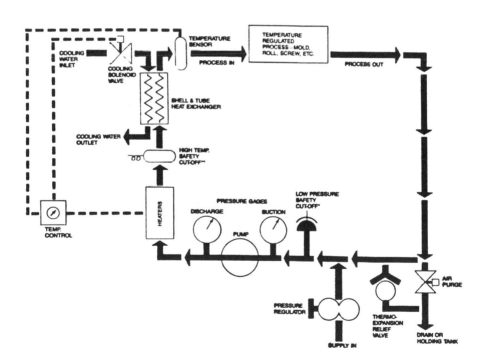

Fig. 9-16 Closed-loop (indirect) cooling circuit. Courtesy of Mokon

The system essentially creates a vacuum to seal the leak, so there is no need to shut down the process for mechanical repair. Figure 9-18 illustrates how negative pressure can be used in a water temperature control system.

Circulating Oil Temperature Controllers

Some of the new thermoplastic and thermosetting plastic materials require melt temperatures over 600 °F and mold temperatures over 350 °F. Additionally, many mold designers and builders want mold temperature to be control-

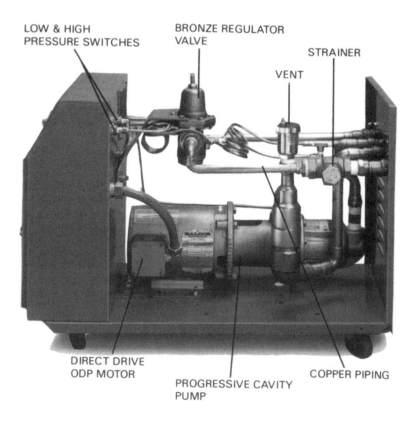

Fig. 9-17 Negative-pressure-based circulating water temperature control system. Courtesy of Mokon

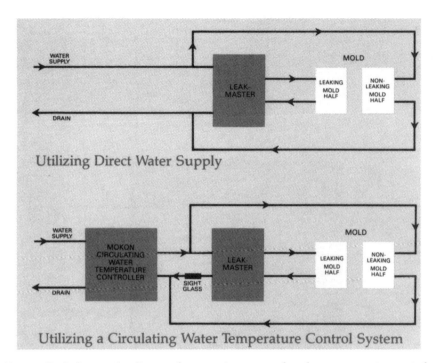

Fig. 9-18 Typical connection diagrams for a negative-pressure-based water temperature control system. Courtesy of Mokon

led by a liquid. These developments have created the need for circulating oil temperature control systems (Fig. 9-19). Although similar in appearance, these systems should not be confused with circulating water temperature controllers. The oil-based systems operate at higher temperatures, and heat exchangers that use water are sometimes needed to help stabilize and ultimately cool down the oil.

Oil-based temperature controllers must be closed-loop in design and must prevent any oil from being exposed to the atmosphere. The oil is actually a heat transfer fluid specifically formulated for such applications. Prior to using oil as a heating/cooling medium, the processor must be sure to have all the correct fitting and hose material for the processes. Hot oil *does not* use the same

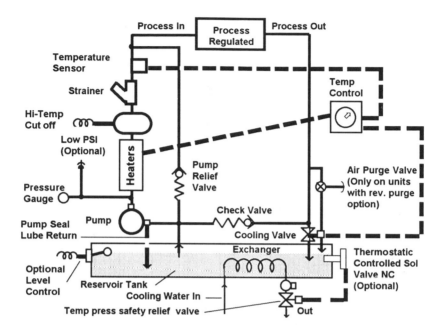

Fig. 9-19 Circulating oil temperature control system. Courtesy of Mokon

hoses and fittings used in circulating water systems. Additionally, operators and technicians should be properly trained in the use and set-up of oil systems prior to installation.

10

Tooling for Plastics Processing

Tooling for plastics processing defines the shape of the part. It falls into two major categories, molds and dies.

A mold is used to form a complete three-dimensional plastic part. The plastics processes that use molds are:

- Injection molding
- Blow molding
- Compression molding
- Reaction injection molding
- Thermoforming

A die is used to form two of the three dimensions of a plastic part. The third dimension, usually thickness or length, is controlled by other process variables. The plastics processes that use dies are:

- Extrusion
- Pultrusion
- Thermoforming

- Lay-up (fiberglass)
- Stamping

Many plastics processors do not differentiate between the terms *mold* and *die*. Molds are the most predominant form of plastics tooling, and unless specified otherwise, the discussion in this chapter refers to molds.

Tool Design

The design of the tooling to produce a specific plastic part must be considered during the design of the part itself. In the past, part designers were not expected to understand plastics processing, tooling, and material. This lack of knowledge usually resulted in the need for several tooling changes prior to processing. Today, most plastics tool designers are as knowledgeable about the capabilities of plastics processes and materials as about the building of molds and dies.

The tool designer must consider several factors that will affect the plastic part:

- The plastic material
- Shrinkage
- Process equipment

Additionally, competitive pressures within the plastics industry require the tool designer to consider how to:

- Facilitate quick tool changeovers
- Optimize tool maintenance
- Simplify (or eliminate) secondary operations
- Ensure product traceability (e.g., date codes and molded-in symbols)

Toolmaking. Historically, plastic molds and dies were built by toolmakers who spent their lives learning and perfecting their craft. It was common practice to have one master toolmaker build the entire mold or die, performing all the machining operations. The quality of the tooling was thus a direct function of the skill of the toolmaker. Today, the average skilled toolmaker is over 55 years old, and very few young people are electing to pursue a career in toolmaking, in the classic sense, because of the time it takes to learn the craft.

The development of numerically controlled (NC) machining centers, computer-based numerically controlled (CNC) machining centers, and computer-aided design systems is filling the void created by the waning numbers of classically trained toolmakers. Molds and dies can now be machined on computer-controlled mills, lathes, and electric discharge machines that require understanding of computers and design, rather than years of broad machining skills. This makes toolmaking more productive, more consistent, and more attractive to younger people considering a trade. In addition, the various machining operations can be assigned to multiple specialists, as opposed to a master toolmaker. As long as the tool design and NC program are accurate, a toolmaker can produce components for several molds or dies in an order that maximizes productivity. The quality of the tool components is now more a function of the equipment than of the toolmaker's skill.

Tool Cost. Fifteen years ago, design and engineering represented about 10% of the cost of a typical mold, materials 10%, and machining labor 80%. Today the trend is that design and engineering represent about 15% of the cost, materials 10%, and machining labor 75%. These proportions reflect the current emphasis on engineering and computer-aided engineering that was not available in the past.

The total cost of a mold can range from less than $1000 for a low-production prototype to well over $500,000 for a high-production mold. Because of these high costs and the fact that many production molds are built under extreme time constraints, there is no room for trial and error. Prototyping has been widely used to evaluate smaller part designs when circumstances and time allow; however, prototyping is not always feasible for larger part designs. There are several alternatives to prototyping:

- *Computer-aided engineering* allows a tool designer to work with a three-dimensional computer model of the tool being designed and to analyze the design.
- *Finite element analysis* allows the tool to be evaluated for production worthiness, on a computer.
- *Rapid prototyping* is a new method of producing a plastic part (or a representative of a plastic part) by using a three-dimensional computer drawing. Sophisticated prototyping apparatus interprets the drawing and guides an articulating laser beam across a specific medium such as a photopolymer, plastic, or laminated paper. The result is a physical representation of the computer-based drawing. Prototyped parts can

be produced in less than 24 hr, and part designs can be scaled to fit the size of the prototyping equipment.

Tool Verification. A tool must be verified to ensure that it will produce the part or product intended at the desired rate and to the design specifications. The responsibility for verifying tool capability is a negotiated component of the purchase order. Most quality toolmakers have the capability to perform their own tool verifications.

For example, the maker of an injection mold might arrange for the mold to be sampled in a molding machine at the processor's plant or in a molding machine similar to that of the processor. The production conditions will be optimized in terms of using the best plastic molding compound, using the best set-up technician, and establishing a stable process that emulates a production process. After the process stabilizes, a number of consecutive samples will be retained for later inspection, usually in two categories:

• *100% inspection* of all dimensions of all cavities, based on one sample of a good shot

• A *statistical machine capability study* of all critical dimensions of all cavities, based on 50 to 100 samples. The total number of critical dimensions should be less than ten. A capability study will reveal any problems associated with a process rhythm that were not apparent in the inspection of a few samples. Also, a capability study recognizes that each cavity is, statistically, a different event (e.g., even if one cavity of a four-cavity mold is within specification, there is no guarantee that all the other cavities are within specification).

Although first-time acceptance of a tool is occurring more frequently, too often a tool must be modified to correct or adjust dimensions, or a customer may change the design of the part after seeing the first samples. Regardless of the reason, the tool design or the part design may be altered once or several times prior to production. The design-to-production flow is illustrated in Fig. 10-1.

Types of Molds

The basic types of molds, regardless of whether they are injection, compression, transfer, or even blow molds, are usually classified by the type and

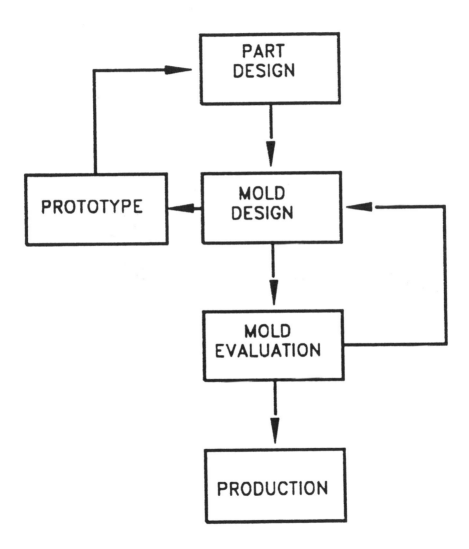

Fig. 10-1 Design-to-production flow

number of cavities they have. For example, Fig. 10-2 illustrates three mold types:

- *Dedicated multiple-cavity mold* (Fig. 10-2a): Each cavity produces the same part. This type of mold is very popular because it is easy to balance the plastic flow and establish a controlled process.
- *Family multiple-cavity mold* (Fig. 10-2b): Each cavity may produce a different part. Historically, family mold designs were avoided because of difficulty in filling uniformly; however, recent advances in mold

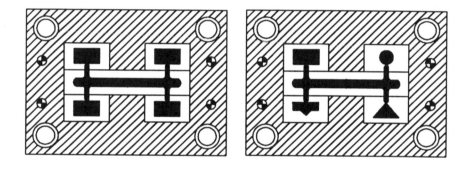

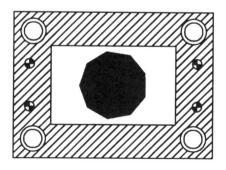

Fig. 10-2 Three basic types of molds. (a) Dedicated multiple-cavity mold. (b) Family mold. (c) Single-cavity mold

building and gating technology make family molds appealing. This is often the case when a processor has a multiple-part assembly and would like to keep inventories balanced.
- *Single-cavity mold* (Fig. 10-2c): One of the simplest mold concepts, this design lends itself to low-volume production and to large plastic part designs.

Injection Molds

Cold Runner Molds. A cold runner mold is the most common type of injection mold design. It commonly has a sprue and runner system, and the resulting shot requires that the plastic parts be separated from the nonproductive runner system. Figure 10-3 shows a cold runner mold and highlights the relationship of the ejector pins to the plastic parts.

Runnerless Molds. The greatest improvement in injection mold technology has been the development of runnerless molds, also called hot runner molds (Fig. 10-4). They incorporate a manifold system, located within the stationary side of the mold, which consists of heated distribution tubes that keep the plastic at a melt consistency.

Runnerless molds allow a plastics processor to eliminate the nonproductive sprue and runner system, and they allow the mold designer to locate the gate on top of the molded part, as opposed to using a parting line or subgates. The gate systems vary, depending on the mold maker's vision of a runnerless system, but most have some common features. As the plastic melt completes its journey through the distribution tubes, it enters a heated probe that continues to keep the plastic in the melt stage. Within the heated probe is a mechanism that allows the probe gate to open or close as a function of the injection pressure or an independent logic system. This gating technique allows the plastics processor to fine-tune the mold fill or to turn a gate off easily if that particular part is not required.

Stack Molds. In the past, when a plastics processor wished to increase productivity, additional molding machines were usually purchased. The direct relationship between productivity and the number of molding machines is now being challenged with the development and refinement of stack molds (Fig. 10-5).

The concept of the stack mold is simple, but the tooling is very complicated. A stack mold requires a special variation of the injection molding machine, one that has three platens that create two daylight areas for mounting molds. (The middle platen is often only a frame into which the middle section of the stack

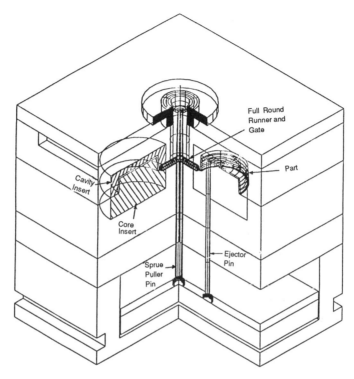

Fig. 10-3 Cutaway view of a cold runner mold, showing the relationship of the ejector pins to the plastic parts

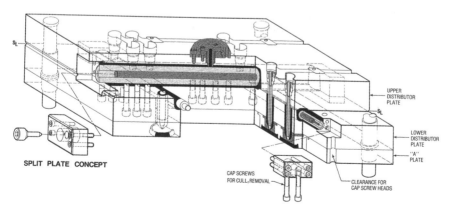

Fig. 10-4 Runnerless mold (hot runner mold). Courtesy of DME

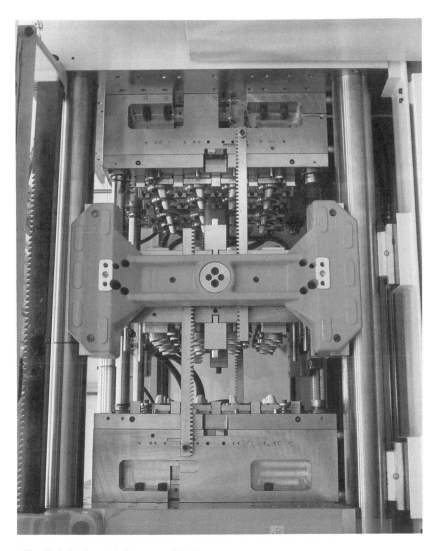

Fig. 10-5 Stack mold. Courtesy of Husky

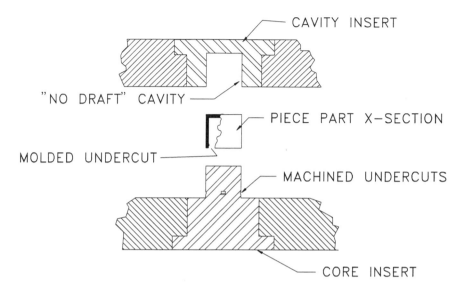

Fig. 10-6 Force "grabbers" that enable a plastic part to be forced off the mold

mold fits.) When the mold is closed, the plastic melt is injected, but instead of being distributed through the stationary mold half, it is distributed through the middle mold section and fills cavities on both sides of the middle section.

Specialty Injection Molds. Many times the design of a part does not lend itself to being molded using conventional molds. Specialty molds are used to perform unique tasks not possible in conventional molds, such as:

- *Unscrewing molds:* used to mold-in threads
- *Two-color molds:* used to mold two or more different plastics at the same time
- *Insert molds:* used to overmold, or insert items in plastic
- *Decorating molds:* allow parts to be decorated while in the mold

One of the terms that must be understood when referring to specialty molds is *undercut.* An undercut is a depression or projection in the molded part that is not perpendicular to the parting line of the mold. Therefore, in order for the plastic part to be ejected, either it must be forced off the mold (Fig. 10-6), or the effect of the undercut must be eliminated by having sections of the mold slide away (Fig. 10-7).

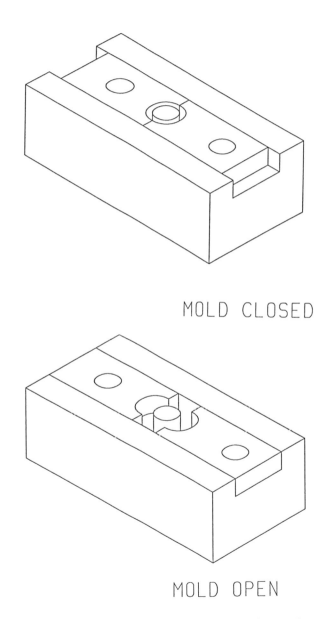

MOLD CLOSED

MOLD OPEN

Fig. 10-7 Sections of the mold may slide away so the plastic part can be ejected

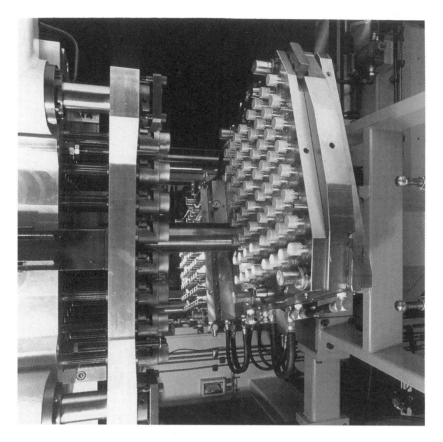

Fig. 10-8 Unscrewing mold

Molded-in threads are undercuts, and unscrewing molds must provide for plastic parts to be turned, so that the plastic part is unscrewed from the mold (Fig. 10-8). These molds are complicated and expensive, and the cost is proportional to the complexity of the molded-in thread.

Blow Molds

Blow molds were relatively simple in design and construction when the predominant blow molded products were bottles and containers. Today, whether for extrusion-based or injection-based blow molding, a blow mold can be as complicated as an injection mold and used to produce nonsymmetrical parts such as fuel tanks or air ducts (see Chapter 5).

Regardless of the degree of sophistication in design, most blow molds:

- Are made of aluminum or other lightweight metals
- Can be either machined or cast
- Experience relative low (under 20 tons) clamping pressure
- Require a high degree of cooling throughout the entire mold
- Require that several vents be located on the parting line and sometimes within the cavity

Mold Bases

The standardized mold base (sometimes referred to as a mold frame) has had a significant impact on plastics tooling productivity. Today companies such as DME, National, Columbia, and Master Unit Die offer a wide variety of standard mold bases and mold components that reduce the time and cost involved in building a new mold.

The sizes of standard mold bases are limited to the sizes of molds that are in the greatest demand. Bases for large molds (over 36 by 36 in.) must be custom-made by the mold maker. Figures 10-9 and 10-10 illustrate typical compression and injection mold bases, respectively.

The mold components and the names of mold details may vary, based on the mold maker's geographic location and experience, but Fig. 10-11 and 10-12 illustrate some of the more common nomenclature for an injection mold base:

- *The locating ring* is a precision-machined ring with an outside diameter of 3.990 in. It is secured to the stationary side of the mold and is used to properly position it onto the mold press platens by fitting into a 4.000 in. diameter hole in the stationary platen.
- *The cavity support plate* provides the foundation of support for the cavity retainer plate and is used to clamp the mold to the stationary platen of the mold press. It is also called the cavity back-up plate or stationary clamping plate.
- *The cavity retainer plate* is the portion of the mold base that usually has machined pockets to hold the cavity inserts.

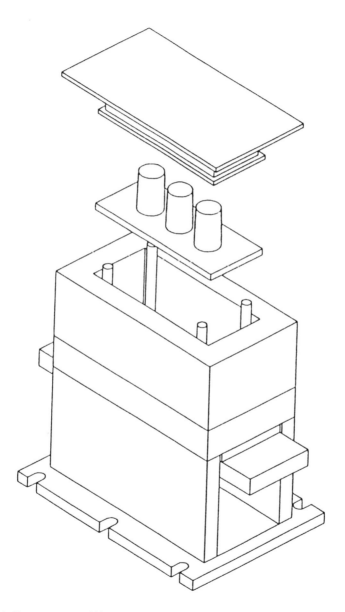

Fig. 10-9 Compression mold base

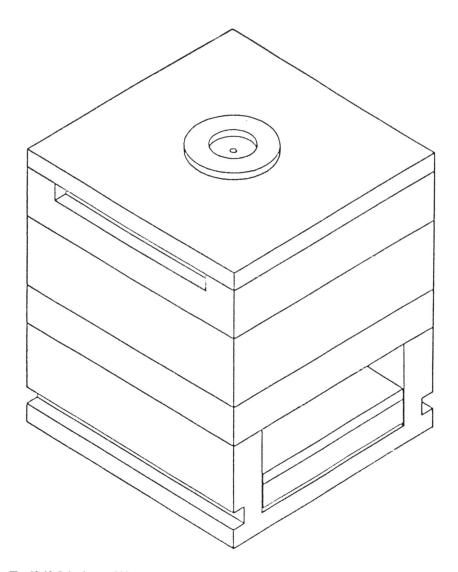

Fig. 10-10 Injection mold base

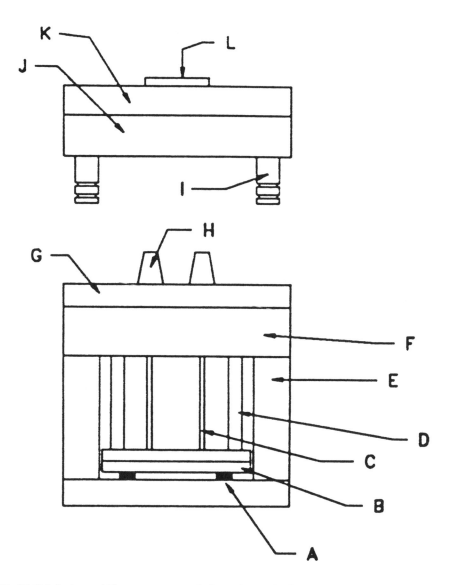

Fig. 10-11 Injection mold base components. A, Spacer button. B, Ejector back-up plate. C, Ejector pin. D, Knockout return pin. E, Parallel or riser. F, Core support plate. G, Core retainer plate. H, Core. I, Guide or leader pin. J, Cavity retainer plate. K, Cavity support plate. L, Locating ring

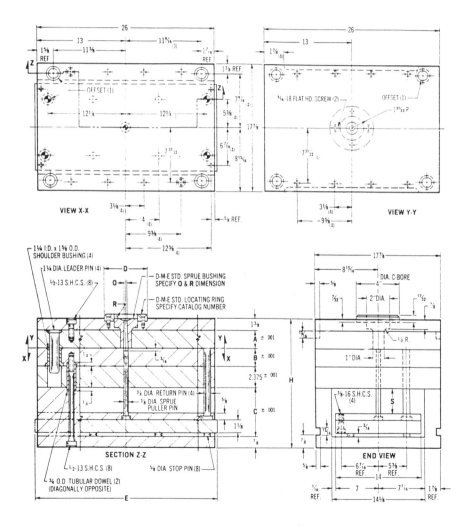

Fig. 10-12 Dimensions of injection mold base. Courtesy of DME

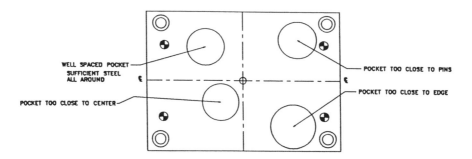

Fig. 10-13 Retainer plate layout

- *The core retainer plate,* like the cavity retainer plate, is usually machined with pockets to hold the core inserts. This plate is also called the force retainer plate.
- *The core support plate* provides the foundation for the core retainer plate, to prevent mold distortion under pressure.
- *Parallels* form the ejector shoe (sometimes called the ejector box), providing the clearance for the mold's ejector mechanism to function. The two parallels may be separate plates or a one-piece system that includes the bottom plate (called the movable clamp plate). Within the ejector shoe resides the ejector system, which consists of the components that make up the remainder of this list.
- *Ejector pins,* also called knockout pins or KO pins, are hard, small-diameter, precision-machined pins that are strategically positioned to eject the plastic part off the core side of the mold.
- *Knockout return pins* do not touch the plastic part. They are dimensioned so that they will touch the face of the opposite (stationary) side of the mold during mold closing. This allows them to be pushed rearward, thus moving the ejector system rearward and allowing the fragile ejector pins to move out of the way when the mold closes.
- *The ejector retainer plate* positions and holds the ejector pins (Fig. 10-13).
- *The ejector back-up plate* supports and secures the ejector retainer plates.
- *The spacer button* keeps the ejector system from making direct contact with the clamp plate. This allows the ejector system to seat properly if there is particulate material present.

- *The parting line* is where the mold halves separate.
- *The witness line* is formed where two or more mold components come together on a surface that will be in contact with plastic material.

The location of the cavity and core inserts within the mold base is an important decision. The mold designer and mold maker must consider the following:

- *Location of mold base components* (bolts, pins, bushings)
- *Distance from the center of the mold,* which determines the runner length required
- *Automation:* Will a robot be readily able to remove the part?

The desire by plastics processors to quickly remove one mold and set up another has existed throughout the history of plastics part manufacturing. The standard mold base, used throughout the plastics industry, requires the mold set-up person to change the entire mold base. The unit mold base permits changing production from one part to another by changing only the mold components (Fig. 10-14). It allows a processor to slide the mold component out of the mold base and leave the bulky mold base clamped within the mold press.

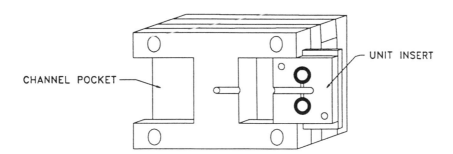

Fig. 10-14 Unit mold base

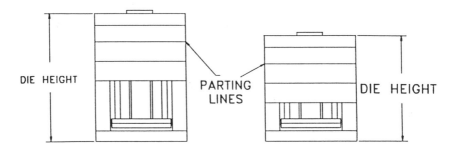

Fig. 10-15 Mold base die height

Mold Design

Die Height. The sum of the stack-up of all the mold base plates (not including the locating ring) is called the die height (Fig. 10-15). The die height is an important mold variable that must be appropriately matched to the specific mold press (see Chapter 6). Die height can be changed by varying the size of the parallels or by adding bolster plates to either or both clamping plates. (A different die height may be required to allow a mold base to fit into a mold press whose daylight specification is beyond the die height of the base.)

Steels. The mold base and the cavity and core components do not have to be produced using the same grade of steel. The mold base requires a steel that can withstand the rigors of the high clamping forces, but the steel does not have to be heat treated beyond nominal hardness. Typical steels for mold bases include SAE 1020 medium-carbon steel and AISI 4130 steel.

The cavity and core steels will be exposed to high injection pressures and to erosive/corrosive plastic materials. Accordingly, it is common practice to heat treat the cavity and core inserts. Typical tool steels for the cavity and core details include A-2, A-6, P-20, and S-7. These tool steels can be heat treated to bring their hardness to the proper level. Other steels can be used to produce cavities and cores that are used in special environments. Stainless steel 420 is used when mold components will be exposed to potentially corrosive plastic materials, such as PVC, and D-2 steel is often used for mold components that will be exposed to highly erosive materials, such as glass-reinforced plastics. D-series steel is one of several high-chromium steels that can be heat treated to levels above RC 60.

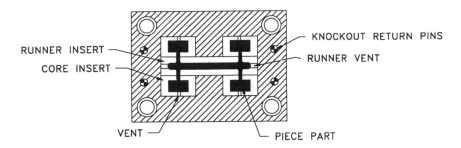

Fig. 10-16 Mold relief area in a four-cavity dedicated mold

Kiss-Off Area. Figure 10-16 highlights the projected area of a mold (shown in black). This is the area where plastic material is under high pressure from the injection forces and is trying to counter the clamp closing force. Directly around the cavity/ core/ runner area is the kiss-off area (shown in white), where the stationary and moving mold halves make contact when the mold is closed. The kiss-off area, sometimes referred to as the masking area, must form a tight seal when the mold is clamped closed. Any irregularity or damage to the kiss-off area will result in a flash condition.

Mold Relief Area. Again referring to Fig. 10-16, the area outside the kiss-off area is called the mold relief area (shown in cross-hatches). This portion of the mold base does not contact plastic material and does not have to kiss off with the other side of the mold during clamping. Mold makers relieve or grind down this area to focus the clamp forces on the kiss-off area, therefore reducing the risk that the mold base will rob precious clamping forces from the kiss-off area.

Figure 10-17 illustrates the relief area and gives a close-up view of how the effectiveness of the kiss-off area can be improved. Most mold makers do not relieve the area directly around the mold base guide pins and bushings, so that the mold will be able to clamp in a stable manner without rocking on the cavity/ core inserts. Relieving the mold base also allows the mold maker and mold designer to vent the cavities more effectively. The vent goes from the cavity to the relief area, not all the way to the outer edge of the mold base itself.

Materials Shrinkage. When a plastic material is being processed, it has a lower density in its melt stage. Also, during processing, the plastic expands when heated and contracts when cooled. These factors, plus the fact that the

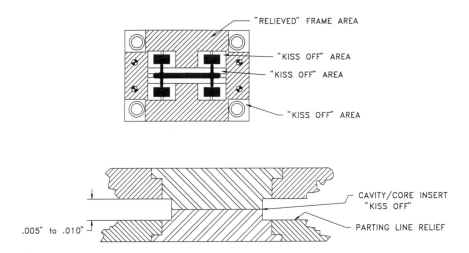

Fig. 10-17   (a) Frame relief and parting line. (b) Parting line relief and kiss-off area

plastic is in an environment of changing pressure, contribute to shrinkage, which affects virtually all plastic materials. Glass-filled plastics shrink less than unfilled plastics, however, and amorphous plastics shrink less than semicrystalline plastics. Semicrystalline and glass-filled plastics are subjected to molecular or fiber orientation that results in an anisotropic shrinkage rate.

Materials shrinkage must be considered by the mold designer and mold maker if the plastic part is to meet specifications. Figure 10-18 shows the basic considerations. Mold makers tend to build a mold on the conservative side of the dimensional tolerances allowed, so they can remove metal if more shrinkage is required. A mold maker will always say, "It's easier to remove metal than to add metal."

Gates. The gate is the mold detail that allows plastic to enter the cavity. Once the cavity has been filled and packed to a desired pressure, the gate must then solidify or freeze off. Historically, gate designs and sizes were determined solely by the experience of the mold designer and mold maker, which often resulted in filling and packing problems. Today, the mold designer has sophisticated computer-based modeling tools and can model the mold and gates prior to the cutting of any steel. This allows corrections and adjustments to be

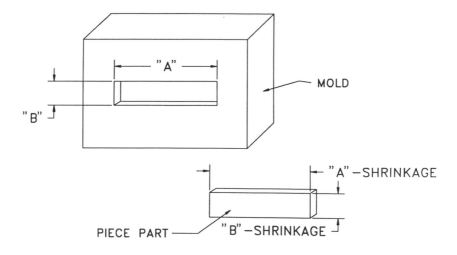

Fig. 10-18 Mold shrinkage

made in a cost-effective manner. There are dozens of gate designs, but a few of the most common are illustrated in Fig. 10-19.

Vents. Vents are usually required in mold cavities, because two things cannot occupy the same space at the same time. When mold cavities are not full of plastic, they are full of air. This air must be displaced if the mold is to fill and pack properly. If the air is not properly removed, the part may have a burn mark, caused by rapid compression heating of the air within the cavity, or it may be underfilled (a "short shot").

Figure 10-20 illustrates a simple cavity and vent. The dimensions of the vent depend on the melt viscosity of the plastic. The vent should be large enough to allow the trapped air to escape, but small enough to prevent any plastic flash from occurring. A typical vent may be 0.100 to 0.200 in. wide and 0.0015 in. deep. The vent or vents are often not specified by the mold designer, because the size, location, and quantity of vents can vary with the mold, the material, and the molding process. To get around this issue, mold designers often include the note "Vent after Sampling" on the mold print. This allows the mold maker to add the correct vents based on how the mold fills during the tool proofing phase of development.

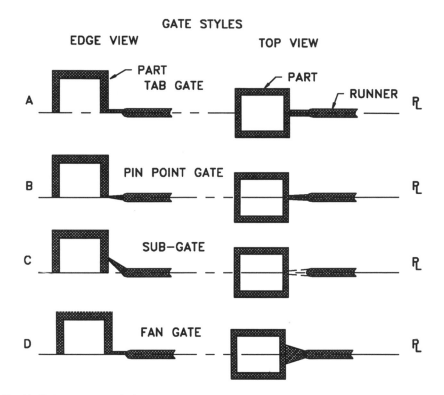

Fig. 10-19 Common gate designs

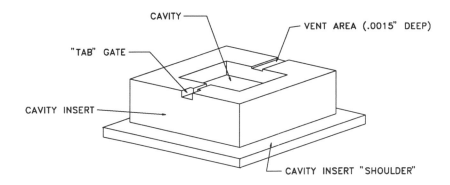

Fig. 10-20 Mold cavity insert with gate and vent

Dies

Within the plastics industry, the term *die* is most often applied to the processes of extrusion (see Chapter 4). Extrusion dies may be categorized by the type of product being produced (e.g., film, sheet, profile, or co-extrusion), but they all have some common features:

- *Steel:* The dies must be made of a high-quality tool steel, hardened so that the areas that contact the plastic material do not erode. (Like the cycling of injection molds, the extrusion process is continuous, making both erosion and corrosion a significant factor.) Additionally, many dies have a dense, hard chrome plating in the area where plastic melt contacts the die.
- *Heaters:* Extrusion dies must be heated to maintain a melt flow condition for the plastic material. Most of the heaters are cartridge-type elements that slip fit into the die at strategic locations. Along with the heaters, the dies have to accommodate temperature sensors, such as thermocouples.
- *Pressure sensors:* Many sophisticated dies are outfitted with sensors that monitor melt pressure. This allows the extrusion processor to better monitor and control the process.
- *Parting line:* Large extrusion dies must be able to separate at the melt flow line for easier fabrication and maintenance. Smaller-profile dies may not have a parting area, as they can be constructed in one piece.
- *Die swell compensation:* In extrusion processes, the extrudate swells when it exits the die, as explained in Chapter 4. Die swell is a function of the type of plastic material, the melt temperature, the melt pressure, and the die configuration. If the die has not been compensated for die swell, the resulting extruded part will not have the correct shape and dimensions.

Figure 10-21 illustrates the basic components of a sheet extrusion die. The die is built in two halves that part for easier construction and maintenance. The use of a jack bolt facilitates separation of the die halves when the die is full of plastic. The die lip can be adjusted (across the entire lip length) to allow the processor to keep the thickness of the sheet within specification.

Figure 10-22 illustrates a T-type die and a coathanger-type die, which are used in both film and sheet extrusion. The die must smooth the plastic melt,

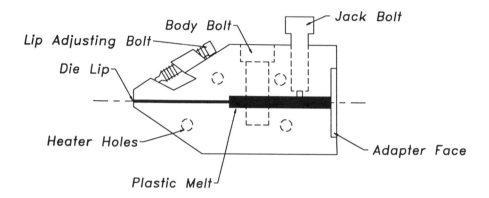

Fig. 10-21 Sheet extrusion die

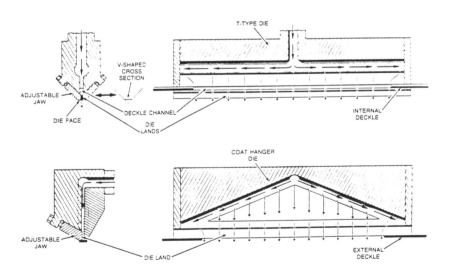

Fig. 10-22 Schematic cross sections of T-type and coathanger-type extrusion dies

which has been rigorously melted and mixed in the extruder, into a uniform laminar flow. The internal shape of the die and the smoothness of the die surface are critical to this successful flow transition. The deckle rods illustrated in Fig. 10-22 are used by the extrusion processor to adjust the width of the extrudate.

Appendix

Length

Physical Quantity		Multiply By	To Convert To	
Centimeters	(cm)	.3937	Inches	(in)
Feet	(ft)	.3048	Meters	(m)
Inches	(in)	2.54	Centimeters	(cm)
Kilometers	(km)	.6215	Miles	(mi)
Meters	(m)	3.281	Feet	(ft)
Miles	(mi)	1.609	Kilometers	(km)
Millimeters	(mm)	.03937	Inches	(in)

Area

Physical Quantity		Multiply By	To Convert To	
Square Centimeters	(cm^2)	.155	Square Inches	(in^2)
Square Feet	(ft^2)	.0929	Square Meters	(m^2)
Square Inches	(in^2)	6.452	Square Centimeters	(cm^2)
Square Meters	(m^2)	10.76	Square Feet	(ft^2)
Square Millimeters	(mm^2)	.00155	Square Inches	(in^2)

Volume

Physical Quantity		Multiply By	To Convert To	
Cubic Centimeters	(cm^3)	.061	Cubic Inches	(in^3)
Cubic Feet	(ft^3)	.02832	Cubic Meters	(m^3)
Cubic Feet	(ft^3)	28316	Cubic Centimeters	(cm^3)
Cubic Feet	(ft^3)	1728	Cubic Inches	(in^3)
Cubic Feet	(ft^3)	7.481	Gallons	(gal)
Cubic Inches	(in^3)	16.387	Cubic Centimeters	(cm^3)
Cubic Meters	(m^3)	1000	Liters	(l)
Cubic Meters	(m^3)	264.18	Gallons	(gal)
Cubic Meters	(m^3)	35.31	Cubic Feet	(ft^3)
Cubic Millimeters	(mm^3)	61×10^{-6}	Cubic Inches	(in^3)
Gallons	(gal)	231	Cubic Inches	(in^3)
Gallons	(gal)	3.7854	Liters	(l)
Liters	(l)	.2642	Gallons	(gal)
Liters	(l)	1000	Milliliters	(ml)
Milliliters	(ml)	.061	Cubic Inches	(in^3)

Density

Physical Quantity		Multiply By	To Convert To	
Grams/Cubic Centimeter	(g/cm^3)	1000	Kilograms/CubicMeter	(kg/m^3)
Grams/Cubic Centimeter	(g/cm^3)	.03613	Pounds/Cubic Inch	(lb/in^3)
Kilograms/Cubic Meter	(kg/m^3)	.06243	Pounds/Cubic Foot	(lb/ft^3)
Kilograms/Cubic Meter	(kg/m^3)	.001	Grams/Cubic Centimeter	(g/cm^3)
Pounds/Cubic Foot	(lb/ft^3)	16.018	Kilograms/Cubic Meter	(kg/m^3)
Pounds/Cubic Foot	(lb/ft^3)	.016018	Grams/Cubic Centimeter	(g/cm^3)
Pounds/Cubic Inch	(lb/in^3)	$.5787 \times 10^{-3}$	Pounds/Cubic Foot	(lb/ft^3)
Pounds/Cubic Inch	(lb/in^3)	27.68	Grams/Cubic Centimeter	(g/cm^3)

Speed/Velocity

Physical Quantity		Multiply By	To Convert To	
Centimeters/Second	(cm/s)	.03281	Feet/Second	(fps)
Feet/Minute	(fpm)	.011364	Miles/Hour	(mph)
Feet/Second	(fps)	.3048	Meters/Second	(m/s)
Meters/Second	(m/s)	3.281	Feet/Second	(fps)
Miles/Hour	(mph)	1.609	Kilometers/Hour	(km/hr)
Miles/Hour	(mph)	88	Feet/Minute	(fpm)

Volumetric Flow Rate

Physical Quantity		Multiply By	To Convert To	
Cubic Centimeters/Second	(cm^3/s)	60×10^{-6}	Cubic Meters/Minute	(m^3/min)
Cubic Feet/Hour	(ft^3/hr)	.12468	Gallons/Minute	(gpm)
Cubic Meters/Minute	(m^3/min)	264.18	Gallons/Minute	(gpm)
Gallons/Minute	(gpm)	.06308	Liters/Second	(l/s)
Gallons/Minute	(gpm)	.22713	Cubic Meters/Hour	(m^3/hr)

Mass

Physical Quantity		Multiply By	To Convert To	
Gram	(g)	.03527	Ounces	(oz)
Kilogram	(kg)	2.205	Pounds	(lb)
Metric Ton		1000	Kilograms	(kg)
Metric Ton		2205	Pounds	(lb)
Ounces	(oz)	28.35	Grams	(g)
Pound	(lb)	.4536	Kilograms	(kg)
Pound	(lb)	453.6	Grams	(g)
Ton		907.18	Kilograms	(kg)

Pressure

Physical Quantity		Multiply By	To Convert To	
Atmospheres	(atm)	1.01325	Bar	
Atmospheres	(atm)	10,332.3	Kilograms/Square Meter	(kg/m^2)
Atmospheres	(atm)	101,325	Kilopascals	(KPa)
Atmospheres	(atm)	14.696	Pounds/Square Inch	(psi)
Atmospheres	(atm)	29.921	Inches Mercury	(in Hg)
Atmospheres	(atm)	33.934	Feet of Water	(ft H_2O)
Atmospheres	(atm)	760	Millimeters Mercury	(mm Hg)
Atmospheres	(atm)	760	Torr	
Bar		14.504	Pounds/Square Inch	(psi)
Pounds/Square Inch	(psi)	703.07	Kilograms/Square Meter	(kg/m^2)
Pounds/Square Inch	(psi)	.0703	Kilograms/Square Centimeter	(kg/cm^2)

Energy

Physical Quantity		Multiply By	To Convert To	
British Thermal Units	(BTU)	.2931	Watts-Hours	(WH)
British Thermal Units	(BTU)	$.393 \times 10^{-3}$	Horsepower-Hours	(Hp-hr)
British Thermal Units	(BTU)	1055	Joules	(J)
British Thermal Units	(BTU)	778.2	Foot-Pounds	(ft-lb)
Calories	(cal)	4.187	Joules	(J)
Foot-Pounds	(ft-lb)	$.3766 \times 10^{-3}$	Watt-Hours	(WH)
Foot-Pounds	(ft-lb)	$.505 \times 10^{-6}$	Horsepower-Hours	(Hp-hr)
Foot-Pounds	(ft-lb)	.001285	British Thermal Units	(BTU)
Foot-Pounds	(ft-lb)	1.3558	Joules	(J)
Horsepower-Hours	(Hp-hr)	.7457	Kilowatt-Hours	(KWH)
Horsepower-Hours	(Hp-hr)	1.98×10^6	Foot-Pounds	(ft-lb)
Horsepower-Hours	(Hp-hr)	2.6845×10^6	Joules	(J)
Horsepower-Hours	(Hp-hr)	2544.4	British Thermal Units	(BTU)
Joules	(J)	$.2778 \times 10^{-3}$	Watt-Hours	(WH)
Joules	(J)	$.3725 \times 10^{-6}$	Horsepower-Hours	(Hp-hr)
Joules	(J)	.7376	Foot-Pounds	(ft-lb)
Joules	(J)	$.948 \times 10^{-3}$	British Thermal Units	(BTU)
Kilowatt-Hours	(KWH)	3412	British Thermal Units	(BTU)
Kilowatt-Hours	(KWH)	1.341	Horsepower-Hours	(Hp-hr)
Kilowatt-Hours	(KWH)	1000	Watt-Hours	(WH)
Kilowatt-Hours	(KWH)	2.655×10^6	Foot-Pounds	(ft-lb)
Kilowatt-Hours	(KWH)	3.6×10^6	Joules	(J)
Watt-Hours	(WH)	.001341	Horsepower-Hours	(Hp-hr)
Watt-Hours	(WH)	2655	Foot-Pounds	(ft-lb)
Watt-Hours	(WH)	3.412	British Thermal Units	(BTU)
Watt-Hours	(WH)	3600	Joules	(J)

Power

Physical Quantity		Multiply by	To Convert To	
British Thermal Units/Hour	(BTU/hr)	$.393 \times 10^{-3}$	Horsepower	(Hp)
British Thermal Units/Hour	(BTU/hr)	.2931	Joules/Second	(J/s)
British Thermal Units/Hour	(BTU/hr)	.2931	Watts	(W)
Horsepower	(Hp)	.746	Kilowatts	(KW)
Horsepower	(Hp)	2544.4	British Thermal Units/Hour	(BTU/hr)
Horsepower	(Hp)	746	Joules/Second	(J/s)
Joules/Second	(J/s)	1	Watts	(W)
Joules/Second	(J/s)	.001341	Horsepower	(Hp)
Joules/Second	(J/s)	3.412	British Thermal Units/Hour	(BTU/hr)
Kilowatts	(KW)	1.341	Horsepower	(Hp)
Kilowatts	(KW)	1000	Joules/Second	(J/s)
Kilowatts	(KW)	3412	British Thermal Units/Hour	(BTU/hr)
Watts	(W)	.001	Kilowatts	(KW)
Watts	(W)	.001341	Horsepower	(Hp)
Watts	(W)	1	Joules/Second	(J/s)
Watts	(W)	3.412	British Thermal Units/Hour	(BTU/hr)

Selected Reference Sources

- ASM International, Materials Park, OH
- Attwood Corporation, 1016 N. Monroe St., Lowell, MI
- Battenfeld of America, Inc., 31 James P. Murphy Industrial Highway, W. Warwick, RI
- Bekum Plastics Machinery, Inc., 1140 W. Grand River, Williamston, MI
- Brown Machine Company, Beaverton, MI
- Cincinnati Milacron Company, Plastics Machinery Systems, 4165 Halfacre Rd., Batavia, OH
- Coastal Engineered Products Company, Varnville, SC
- Crane Plastics, 2141 Fairwood Ave., Columbus, OH
- Eagle-Picher Industries, Inc., Plastic Division, 14123 Roth Rd., Grabill, IN
- Husky Injection Molding Systems Ltd., 530 Queen St. S., Bolton, Ontario, Canada
- Johnson Controls, Manchester, MI

- McGraw-Hill Publishing Company, *Modern Plastics Magazine*, 1221 Ave. of the Americas, New York, NY
- Society of Plastics Engineers, 14 Fairfield Dr., Brookfield, CT
- Society of the Plastics Industry, 1275 K St., N.W., Washington, DC
- Steelcase, Inc., Grand Rapids, MI

Index

material loaders, 256-257, 258, 259, 260
mold temperature controllers, 262, 264
plenum hopper, 242, 244, 245, 247-248, 249, 250-251, 254-255
typical equipment listed, 241
Anti-reset, 59
Antistatic agents, 21, 40-41
external, 40, 41
internal, 40-41
ion discharge, 40, 41
APC-2, supplier, resin, fiber type, sheet structure, and temperatures, 194
Archimedian screw, 85
Area, conversion table, 299
Autoclave/vacuum forming, methods and equipment, 197, 200
Automatic reset (integral), 59
Automatic screen changers, 82

Back pressure, 162
Barrel thermoset injection molding machine, 170
Barrier plastics, 103
Bearing shank, 74
Benzene, 29
Beryllium-copper
for blow mold material, 136
as tooling material for low-volume thermoform production, 216
Biaxial orientation, 129-132
advantages, 131-132
methods, 130-131
systems, 130
Biocides, 41
Blending, 34, 35
Blow molding, 103-143
equipment, 104, 106
equipment operation, 140-142
extrusion blow molding, 103, 104-124

injection blow molding, 103, 124-132
molds, 132-136, 271
products produced, 105
shutdowns, 143
tooling, 136-140
Blow molds, 282-283
Blown film extrusion, 96-99
Blown flash, 133-134
Blow-up ratio, 116
defined, 109
Boost pressure, 162
Branched polymers, 35-36
Brass, for resin casting molds, 225, 226-227
Breaker plate, 81-83
Bulk packaging, 14
Butadiene, 34

CAB, thermoforming, 219
Captive plastics processors, 5-6
Captured neck system, 116
Carbon black, 38
UV stabilization and, 40
Carbon chain, 27, 28, 31, 36
Carbon/graphite, as reinforcement material for plastics, 37
Castable "steel-like" alloys, material for SRIM molds, 184
Cast aluminum
material for SRIM molds, 184
as tooling material for high-volume production molds for the thermoforming process, 216
Cast film extrusion, 99-101
Cavity back-up plate. *See Cavity support plate*
Cavity retainer plate, 283, 286
Cavity support plate, 283, 286
Central hopper loader, 257, 259